Statistics Level 3

Statistics Level 3

ERIC WALKER, MA(Cantab.)

Formerly Head of the Mathematics Department and Deputy Headmaster of Sir Roger Manwood's School, Sandwich; later at the South Kent College of Technology, Folkestone.

HOLT, RINEHART AND WINSTON
LONDON · NEW YORK · SYDNEY · TORONTO

Holt, Rinehart and Winston Ltd: 1 St Anne's Road,
                                  Eastbourne, East Sussex BN21 3UN

British Library Cataloguing in Publication Data

Walker, Eric
  Statistics Level 3.—(Holt technician texts)
  1. Shop mathematics   2. Mathematics statistics
  I. Title   II. Series
  519.5'0246   TJ1165

ISBN 0–03–910356–0

Typeset by Macmillan India Ltd., Bangalore.
Printed in Great Britain by Mackays of Chatham Ltd, Chatham, Kent.

Last digit is print no: 9 8 7 6 5 4 3 2 1

Preface

The book is designed to cover the syllabus for the standard half unit TEC U80/688.

The electronic calculator has been used extensively throughout the book. At times there has been no attempt to round off intermediate calculations and final results in order that the student using his or her own calculator might follow more readily the various stages of the examples. A similar procedure has been followed in presenting many of the answers to exercises. The student may round off such values to any degree of accuracy desirable in the particular cases.

Certain sections of explanations and certain examples and exercises are more difficult to understand than others. The more difficult passages are marked with either one or two asterisks. Normally the student will find that a section marked by two asterisks is more difficult than a section marked by one asterisk.

A student who intends to pursue the subject to a higher level ought to aim at studying and understanding all parts of the book. A student who does not wish to proceed beyond Level 3 ought to aim at completing all unmarked sections and the majority of those marked with one asterisk. When a marked explanation is found to be too difficult it is almost always essential that the final principle, rule or formula of that section be noted carefully.

Perhaps the two chapters which will be found most difficult to understand are Chapters 5 and 6, especially the latter.

I wish to express my sincere appreciation to my son-in-law, Dr A. N. Godwin, for the valuable discussion which we had about this section and its inherent difficulties, as well as about a suitable format for the chapter on hypothesis testing.

E. WALKER

To my secretary, for her patience and endurance in typing the various drafts and final version, and for the care shown in checking the work at all stages.

Contents

5 Point and Interval Estimates

72

6 Simple Hypothesis Testing

90

7 Basic Ideas of Correlation

101

8 Basic Ideas of Regression

113

Unit Reference

1

Probabilities

1.1 Simple probabilities

When an event, such as the throwing of a die, is due to take place, and a number of outcomes – that the die will fall with one, two, three, four, five or six face up – are likely to take place, we can never be certain that we can predict any outcome before it takes place. Other examples are the event or trial of: tossing a coin, with two possible outcomes; drawing a card from a standard pack of playing cards, with 52 possible outcomes; the weather in your town on a date six weeks ahead, with perhaps four possible outcomes – dry and cold, dry and warm, wet and cold, or wet and warm. In fact, in this case the possible outcomes might be even more elaborate and numerous than the ones indicated, but the ones referred to would suffice to give a broad description of any particular day.

Too many factors influence the result of any such event. In tossing a coin we must take into consideration the height to which it is tossed, the rate of spin as it leaves the hand, the presence of wind or draught, and the nature of the surface onto which the coin falls, to mention but a few. Now think of the number of factors which will influence the result of drawing a card from a pack, of throwing a die, or of determining the weather in your town on a particular day. Even when it is known that the coin is not out of balance (the coin is said to be fair), when the deck of cards has not been unfairly stacked and the die not weighted, the other factors which help to determine the outcome are either so numerous or so obscure that it is impossible to assess their influence with certainty. Even when we think we are in possession of all the relevant facts to make a prediction about, for example, the result of a

1

football match, a boxing contest, a horse race, an athletics event, or an election, the real outcome will often prove our forecast to be wrong, and in the end we may be unaware of what factor or factors upset our prediction.

Consider in more detail three of the events or trials referred to: tossing a fair coin, throwing a fair die, and drawing a card from a standard pack of playing cards. There are two principal ways of approaching the problems. The first is by carrying out a series of trials and carefully tabulating the outcomes. The second is by asking ourselves what results we could reasonably expect to happen.

Tossing a coin

It is probably not too difficult for any of us to toss a coin 50 times and note the number of times it comes down head (H) and the number of times it comes down tail (T). We might even be able to do it 100 times without becoming bored. But beyond that we might lose patience. If you can spare the time, carry out 100 trials. What were your results?

Suppose, without doing it, we think about the event. What results would we expect to occur? That out of 50 trials we would have $25\,H$ and $25\,T$, that out of 100 trials we would have $50\,H$ and $50\,T$, and so on? That might, in fact, be our first impression, but if we put the matter to the test it is highly unlikely that those results would occur. What is certain is that greater the number of trials, n, the more nearly the number of times the outcome was H would approach the value $\frac{1}{2}n$; similarly with the outcome T.

Note. We do not say that if there were 1000 trials the number of tail outcomes, $n(T)$, would be 500 and the number of head outcomes, $n(H)$, would also be 500. In the event of 1000 trials the number of head outcomes might be 483 and the number of tail outcomes 517, i.e. $n(H) + n(T) = 483 + 517 = 1000$.

Note. We *do* say that if N is the total number of trials and $n(H) = n$ then $n(T) = N - n$ so that:

$$n(H) + n(T) = n + N - n = N$$

And we do say that:

$$\frac{n}{N} \to \tfrac{1}{2} \text{ the larger } N \text{ becomes} \qquad\qquad 1.1$$

We also say that:

$$\frac{N - n}{N} \to \tfrac{1}{2} \text{ the larger } N \text{ becomes} \qquad\qquad 1.2$$

1.1 may be expressed alternatively as follows. 'The probability that the outcome is H is $\frac{1}{2}$.' In symbols:

$$P(H) = \tfrac{1}{2} \qquad\qquad 1.3$$

The alternative to *1.2* is: 'The probability that the outcome is T is $\frac{1}{2}$.' In symbols:

$$P(T) = \tfrac{1}{2} \qquad\qquad 1.4$$

Note:

$$P(H) + P(T) = \tfrac{1}{2} + \tfrac{1}{2} = 1$$

Throwing a die

Since there are six possible outcomes to any trial the execution of the trials on a large enough scale and the tabulation of the results would be time consuming and patience exacting. So what do we expect from a fair die? *Note.* We do not say that if there were exactly 600 trials the number of outcome one, i.e. $n(1)$, would be 100, that $n(2) = 100$, and so on. In other words, we do not say that $n(1) = n(2) = n(3) = n(4) = n(5) = n(6) = 100$. The results of 600 actual trials might be: $n(1) = 112$; $n(2) = 105$; $n(3) = 86$; $n(4) = 94$; $n(5) = 108$; $n(6) = 95$.
Note. It is true that:

$$n(1) + n(2) + n(3) + n(4) + n(5) + n(6)$$
$$= 112 + 105 + 86 + 94 + 108 + 95 = 600$$

because there cannot be any outcomes other than 1, 2, 3, 4, 5 or 6. These are the only possibilities.
Note. We do not say that if N is the total number of trials and if $n(1) = n_1$, $n(2) = n_2$, $n(3) = n_3$, $n(4) = n_4$, $n(5) = n_5$ and $n(6) = n_6$ then:

$$n_1 + n_2 + n_3 + n_4 + n_5 + n_6 = N \qquad\qquad 1.5$$

$$\text{and} \quad \frac{n_1}{N} \to \frac{1}{6}, \frac{n_2}{N} \to \frac{1}{6}, \frac{n_3}{N} \to \frac{1}{6} \qquad\qquad 1.6$$

and so on the larger N becomes. *1.6* is expressed alternatively as follows: 'The probability that the outcome is 1 is $\frac{1}{6}$.' In symbols:

$$P(1) = \tfrac{1}{6}$$

In fact:

$$P(1) = P(2) = P(3) = P(4) = P(5) = P(6) = \tfrac{1}{6} \qquad\qquad 1.7$$

Therefore:

$$P(1) + P(2) + P(3) + P(4) + P(5) + P(6) = 6 \times \tfrac{1}{6} = 1 \qquad 1.8$$

Drawing a card from a standard pack

The four suits will be described as S, H, C and D. The different card faces will be described as $A, 2, 3, \ldots 10, J, Q, K$. For a large number of trials, N, suppose the number of times a spade is drawn is $n(S)$, the number of times a three is drawn is $n(3)$, the number of times a jack is drawn is $n(J)$, and so on. Then:

$$\frac{n(S)}{N} \to \frac{13}{52} = \frac{1}{4} \text{ the larger } N \text{ becomes}$$

Note that there are 52 different cards and 13 spades. Again:

$$\frac{n(3)}{N} \to \frac{4}{52} = \frac{1}{13} \text{ the larger } N \text{ becomes}$$

There are 52 different cards and 4 threes. In symbols:

$$P(S) = \tfrac{1}{4}; \ P(3) = \tfrac{1}{13}$$

Excercise 1.1

1. When R represents a red card calculate $P(R)$.
2. When B represents a black card calculate $P(B)$.
3. Calculate: (a) $P(D)$; (b) $P(C)$; (c) $P(H)$.
4. Calculate: (a) $P(H \text{ or } C)$; (b) $P(S \text{ or } D)$; (c) $P(\text{not } D)$.
5. When F represents a face card (jack, queen or king) calculate $P(F)$.
6. When ace counts as high calculate the probability of drawing a card with a value less than six, i.e. $P(< 6)$.
7. For a single die calculate the probability of throwing more than a four, i.e. $P(> 4)$.
8. For a single die calculate the probability of throwing a two or a five, i.e. $P(2 \text{ or } 5)$.
9. For a single die calculate the probability of throwing a one, a three, a four or a six.
10. For a single die calculate the probability of throwing: (a) less than a four; (b) not a four.
11. A bag contains eight red, six white and nine black balls. A ball is drawn at random from the bag. What are the probabilities that the ball drawn is: (a) red, $P(R)$; (b) white, $P(W)$; (c) black, $P(B)$?

12. A bag contains N balls of which n are blue, p are yellow, q are red and the remainder are multicoloured. Calculate the probabilities of: (a) drawing a red ball; (b) drawing a blue or a yellow ball; (c) drawing a multicoloured ball.

The early part of the chapter has assumed certain things:

1. It is equally likely when tossing a fair coin that the result will be H as it will be T.
2. It is equally likely when throwing a fair die that the result will be 1, 2, 3, 4, 5 or 6.
3. It is equally likely when drawing a card from a standard pack of cards that the five of spades will be drawn as it is that the ten of hearts or any other card will be drawn.
4. In tossing a coin the outcomes H and T are *exclusive*, i.e. a coin showing H cannot at the same time show T; similarly that a die showing 1 cannot at the same time show 2, 3, 4, 5 or 6.
5. Most important, we have defined the probability of a particular outcome as follows: when the number of trials of an event is indefinitely large and there are N different possible outcomes, all equally likely, and among those N there are n possible outcomes of a kind A, we write the probability of outcome A as $P(A)$ and say that it equals $\dfrac{n}{N}$ or, if we wish to be more explicit, $\dfrac{n(A)}{N}$.
6. Where the possible outcomes are A, B, C, etc. and A, B, C, etc. are all different from each other, i.e. exclusive, then:

$$n(A) + n(B) + n(C) + \ldots = N \qquad \textit{1.9}$$

where N is the total number of different possible outcomes. Therefore:

$$P(A) + P(B) + P(C) + \ldots = \frac{n(A)}{N} + \frac{n(B)}{N} + \frac{n(C)}{N} + \ldots$$

$$= \frac{N}{N}$$

$$= 1 \qquad \textit{1.10}$$

For example, in drawing a card from a pack:

$$P(S) = \frac{13}{52} = \frac{1}{4}; \ P(H) = \frac{13}{52} = \frac{1}{4}; \ P(C) = \frac{13}{52} = \frac{1}{4}; \ P(D) = \frac{13}{52} = \frac{1}{4}$$

And $\quad P(S) + P(H) + P(C) + P(D) = \frac{1}{4} + \frac{1}{4} + \frac{1}{4} + \frac{1}{4} = 1$

7. In tossing a coin there are no outcomes other than H and T, that is, the outcomes H and T are *exhaustive*. In throwing a die there are no outcomes other than 1, 2, 3, 4, 5 and 6. These outcomes are *exhaustive*. So we can say, in general, where $n(A)$, $n(B)$, $n(C)$, etc. represent the number of different possible outcomes of A, B, C and so on out of a total number of different possible outcomes, N, such that A, B, C, etc. are both exclusive and exhaustive, then:

$$P(A) + P(B) + P(C) + \ldots = 1$$

And, again, where the exclusive outcomes A, B, C, etc. are such that:

$$P(A) + P(B) + P(C) + \ldots = 1$$

then these outcomes are also exhaustive.

Example

The manufacturers of a particular electronic calculator always replace a defective model with a new one. They know from experience that, for every 10 000 of that model sold, 250 will be returned faulty within 12 months of sale and that only 20 will be returned faulty after that period. The remainder are presumed to be fault free, even though faults may have occurred and the buyer has not returned the calculator. What is the probability that this particular model, when bought, can be presumed by the manufacturer to be fault free?

Call $P(F_1)$ the probability that faults develop within one year, $P(F_2)$ the probability that faults appear after the first year, and $P(FF)$ the probability that the model is fault free. Then F_1, F_2 and FF are mutually exclusive and exhaustive outcomes. So:

$$P(F_1) + P(F_2) + P(FF) = 1$$

Now $P(F_1)$ we shall take to be $\frac{250}{10\,000}$ because we shall assume that the data are based on a large number of models sold, and $P(F_2)$ we shall take to be $\frac{20}{10\,000}$ for a similar reason. Then:

$$P(FF) + \frac{250}{10\,000} + \frac{20}{10\,000} = 1$$
$$P(FF) + 0.025 + 0.002 = 1$$
$$P(FF) + 0.027 = 1$$
$$P(FF) = 0.973$$

Exercise 1.2

1. A digit is chosen at random from the following set: 2, 3, 4, 5, 6, 7, 8. Calculate the probabilities that the digit chosen is: (a) even; (b) odd.

*2. The letters A, C, D, H and S are selected to form words with up to five letters. Calculate $P(1)$, $P(2)$, $P(3)$, $P(4)$ and $P(5)$ where $P(n)$ is the probability that n letters selected in order constitute a proper word. Calculate the probability that no matter how many letters are selected, in a given order, the sequence will not represent a real word.

3. Out of a committee of ten men and nine women a subcommittee of two is to be elected. Assume that it is equally likely that any member of the committee will be elected. Calculate the probabilities that the subcommittee will be: (a) two men; (b) two women; (c) a man and a woman.

**4. In preparation for a particular international match 25 England players are selected to form a squad. Of these, five play with Ipswich, two play with Nottingham Forest, three play with Liverpool and four play with Aston Villa. The rest are single players chosen from other clubs. If all players are equally likely to be selected for the final 13, determine the probabilities that: (a) all five Ipswich players; (b) both Nottingham Forest players; (c) all three Liverpool players; (d) all four Villa players; will be in the team, including substitutes. Determine: (e) the probability that the final team choice, including substitutes, will contain no player from any one of the four clubs mentioned; and (f) the probability that at least one player from the four league clubs mentioned will be one of the final 13 chosen.

1.2 Dependent and independent events

If we throw a fair die twice and we call the outcome of the first throw A and outcome of the second throw B then the probability that B is six, which we might write as $P(B = 6)$, is 1/6. It will always be 1/6 no matter what the outcome of the first throw, A, is. In such cases we describe the two events as *independent*. The outcome of A is not influenced by B and that of B is not influenced by A.

Definition

Two events are *independent* when the outcome of one does not influence the outcome of the other. Note that it is sometimes erroneously believed, especially by gamblers, that if the first throw is not a six, then the second throw is more likely to be a six. There is no justification for such a belief.

Definition

Two events are *dependent* when the outcome of one is influenced by the outcome of the other.

Examples

1. Determine the probability that two successive throws of a die will each be six.
    One way of approaching this is to laboriously write down all possible pairs of throws:

$$\begin{array}{cccccc}
1,1 & 1,2 & 1,3 & 1,4 & 1,5 & 1,6 \\
2,1 & 2,2 & 2,3 & 2,4 & 2,5 & 2,6 \\
3,1 & 3,2 & 3,3 & 3,4 & 3,5 & 3,6 \\
4,1 & 4,2 & 4,3 & 4,4 & 4,5 & 4,6 \\
5,1 & 5,2 & 5,3 & 5,4 & 5,5 & 5,6 \\
6,1 & 6,2 & 6,3 & 6,4 & 6,5 & 6,6
\end{array}$$

The only combination to produce the required result is 6,6, i.e. one outcome. The total number of different equally possible outcomes is 36. Then:

$$P(6 \text{ then } 6) = \frac{1}{36} \text{ which might be expressed } \frac{1}{6} \times \frac{1}{6}$$

$$\text{Also } P(2 \text{ then } 3) = \frac{1}{36} \text{ which might be expressed } \frac{1}{6} \times \frac{1}{6}$$

$$\text{And } P(3 \text{ then } 2) = \frac{1}{36} \text{ which might be expressed } \frac{1}{6} \times \frac{1}{6}$$

An alternative approach is to use a diagram such as Fig. 1.1. The only combination 6,6 occurs at the point marked X. The resulting calculations are as above. Where such a procedure is adopted each diagram represents the total number of possible different outcomes, all

Second throw

	1	2	3	4	5	6
1	·	·	·	·	·	·
2	·	·	·	·	·	·
3	·	·	·	·	·	·
4	·	·	·	·	·	·
5	·	·	·	·	·	·
6	·	·	·	·	·	*X*

First throw

Figure 1.1

equally likely. It is referred to as the *total possibility space* or the *sample space*.

2. Determine the probability that in both drawing a card from a pack and tossing a coin, a club and a head will be obtained.

The only combination of club and head is the one marked *X* in Fig. 1.2.

Drawing a card

	S	H	C	D
H	·	·	*X*	·
T	·	·	·	·

Tossing a coin

Figure 1.2

The total number of different equally likely outcomes is eight.

$$P(\text{club and head}) = \frac{1}{8} = \frac{1}{4} \times \frac{1}{2}$$

$$\text{Again,} \quad P(\text{spade and tail}) = \frac{1}{8} = \frac{1}{4} \times \frac{1}{2}$$

$$\text{Further,} \quad P(\text{red card and head}) = \frac{2}{8} = \frac{1}{4} = \frac{1}{2} \times \frac{1}{2}$$

An analysis of the above results produces:

$$P(6 \text{ then } 6) = \frac{1}{6} \times \frac{1}{6} = P(6) \times P(6)$$

$$P(2 \text{ then } 3) = \frac{1}{6} \times \frac{1}{6} = P(2) \times P(3)$$

$$P(3 \text{ then } 2) = \frac{1}{6} \times \frac{1}{6} = P(3) \times P(2)$$

$$P(\text{club and head}) = \frac{1}{4} \times \frac{1}{2} = P(C) \times P(H)$$

$$P(\text{spade and tail}) = \frac{1}{4} \times \frac{1}{2} = P(S) \times P(T)$$

$$P(\text{red card and head}) = \frac{1}{2} \times \frac{1}{2} = P(\text{red card}) \times P(H)$$

These results lead us to believe that if A and B are independent events then:

$$P(A \text{ and } B), \text{ or } P(A \text{ with } B) = P(A).P(B)$$

To prove this result suppose the event A occurs a times in a sample space n. Then not A occurs $(n-a)$ times: A and not A are exclusive and exhaustive. A and not A cover all possibilities.

Suppose that event B occurs b times in a sample space m. Then not B occurs $(m-b)$ times: B and not B are exclusive and exhaustive.

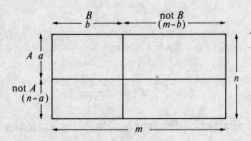

Figure 1.3

A with B occurs ab times (Fig. 1.3). The total sample space $= mn$. Then:

$$P(A \text{ with } B) = \frac{ab}{mn} = \frac{a}{n} \times \frac{b}{m} = P(A) \times P(B)$$

i.e. $P(A \text{ with } B)$ or $P(A \text{ and } B) = P(A).P(B)$ *1.11*

where A and B are independent events.

Note. A and B cannot be independent when they are mutually exclusive. This applies especially when A and B are exhaustive. In that case:

$$P(A) + P(B) = 1$$

$$\text{Therefore} \quad P(B) = 1 - P(A)$$

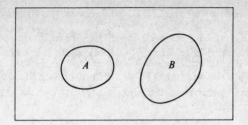

Figure 1.4

The probability of *B* depends on the probability of *A*, so the events *A* and *B* cannot be independent. In general, when the events *A* and *B* are mutually exclusive (Fig. 1.4), *A* and *B* cannot occur simultaneously. Therefore $P(A$ and $B) = 0$, and this must be so even when $P(A) \neq 0$ and $P(B) \neq 0$. And, if $P(A)$ and $P(B)$ were each equal to 0, the case would indeed be trivial. Therefore formula *1.11* cannot apply when *A* and *B* are mutually exclusive events. Formula *1.11* is sometimes called the multiplication of probabilities.

1.3 The addition law of probabilities

Suppose we wish to calculate the probability that either the event *A* occurs or the event *B* occurs, or that they both occur, that is, $P(A$ or $B)$. From Fig. 1.3 we obtain the following.

The number of times that *A* alone occurs $= a(m - b)$. This is extracted from the top right-hand corner rectangle, in other words, when *A* and not *B* occur.

The number of times that *B* alone occurs comes from the bottom left-hand corner rectangle and equals $b(n - a)$. It is represented by *B* and not *A*.

The number of times that they both occur comes from the top left-hand corner rectangle and equals ab. Then:

$$
\begin{aligned}
P(A \text{ or } B) &= \frac{a(m - b) + b(n - a) + ab}{mn} \\
&= \frac{am + bn - ab}{mn} \\
&= \frac{am}{mn} + \frac{bn}{mn} - \frac{ab}{mn} \\
&= \frac{a}{n} + \frac{b}{m} - \frac{ab}{mn} \\
&= P(A) + P(B) - P(A \text{ and } B) \qquad\qquad 1.12
\end{aligned}
$$

Alternative methods of writing *1.11* and *1.12* are the following:

$$P(A \text{ or } B) = P(A + B) = P(A) + P(B) - P(A) \cdot P(B), \text{ for } 1.12$$
$$P(A \text{ and } B) = P(A \cdot B) = P(A) \cdot P(B), \text{ for } 1.11$$

$P(A \text{ and } B)$ may also be written as $P(A \cap B)$ and $P(A \text{ or } B)$ may be written as $P(A \cup B)$, where $\cap$ means conjunction or intersection and $\cup$ means union.

Examples

1. Suppose we draw a card from a standard pack of cards and we also throw a fair die. What is the probability that we shall draw a card higher than or equal to seven (ace counts as high) and throw a number less than five?
    The two events are independent.

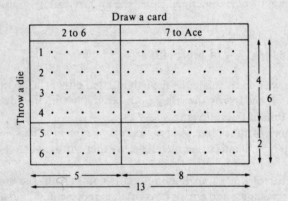

Figure 1.5

From Fig. 1.5, for the card $P(\geqslant 7) = \dfrac{8}{13}$

For the die $P(< 5) = \dfrac{4}{6} = \dfrac{2}{3}$

By the multiplication law of independent events (*1.11*):

$$P(\geqslant 7 \text{ and } < 5) = \frac{8}{13} \times \frac{2}{3} = \frac{16}{39}$$

From the sample space, $n(\geqslant 7 \text{ with } < 5) =$ number in the top right-hand rectangle $= 8 \times 4 = 32$. The total sample space is 13×6.

Therefore:

$$P(\geqslant 7 \text{ with } < 5) = \frac{32}{13 \times 6} = \frac{16}{39}$$

The diagrammatic approach will still be useful in certain problems, so we shall not ignore it. However, for independent events we shall apply formula *1.11*.

Suppose we wish to determine the probability that we shall draw a card higher than or equal to seven or that we shall throw a number less than five with the die, or both.

By formula *1.12*:

$$P(\geqslant 7 \text{ or } < 5) = P(\geqslant 7) + P(< 5) - P(\geqslant 7 \text{ with } < 5)$$
$$= \frac{8}{13} + \frac{2}{3} - \frac{16}{39}$$
$$= \frac{24}{39} + \frac{26}{39} - \frac{16}{39} = \frac{34}{39}$$

2. If two dice are thrown at the same time what is the probability that neither will show a one or a two?

   Suppose that outcome *A* represents a die which shows a one or a two. Then the outcome $\sim A$, or not *A*, represents a die which shows anything but a one or a two, i.e. a three, a four, a five or a six. Then, for any fair die:

   $$P(A) = \frac{2}{6} = \frac{1}{3} \text{ and } P(\sim A) = \frac{2}{3}$$

   The outcome *A* for any die is not influenced by the outcome for any other die. The events are independent.

   An alternative diagrammatic method is to use a tree diagram, as in Fig. 1.6. The probability that neither shows a one or a two is the probability that both do not show a one or a two, that is:

   $$P(\sim A \text{ and } \sim A) = 4/9$$

   The probability that either one or the other die shows a one or a two but not both is:

   $$P(A \text{ and } \sim A) + P(\sim A \text{ and } A)$$
   $$= \frac{2}{9} + \frac{2}{9} = \frac{4}{9}$$

3. What is the probability of drawing in succession two diamonds from a pack of cards, each card being replaced in the pack?

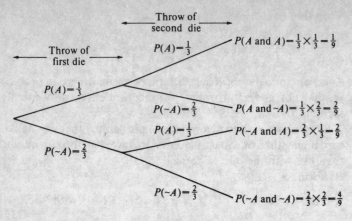

Figure 1.6

Suppose outcome A means the first card drawn is a diamond. Suppose outcome B means the second card drawn is a diamond. Then $P(AB)$ is taken to mean the probability that both cards drawn are diamonds. From Fig. 1.7:

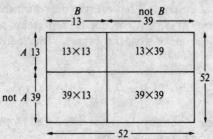

Figure 1.7

$$P(AB) = \frac{n(AB)}{N} = \frac{13 \times 13}{52 \times 52} = \frac{1}{16}$$

or, since A and B are independent:

$$P(AB) = P(A) \times P(B) = \frac{13}{52} \times \frac{13}{52} = \frac{1}{4} \times \frac{1}{4} = \frac{1}{16}$$

*4. A bag contains 15 balls, eight of them red and the remainder white. Two balls are picked out in succession and returned to the bag. Determine the probabilities of picking out: (a) two white balls; (b) one red and one white; (c) two red balls.

Suppose the probability of picking out two white balls is $P(WW)$; of picking out one red and one white is $P(RW)$; and of picking out two red is $P(RR)$. Since the events are independent, then:

(a) $P(WW) = P(W) \times P(W) = \dfrac{7}{15} \times \dfrac{7}{15} = \dfrac{49}{225}$

(b) $P(RW) = P(\text{first } R, \text{ then } W) + P(\text{first } W, \text{ then } R)$
$= P(R) \times P(W) + P(W) \times P(R)$
$= \dfrac{8}{15} \times \dfrac{7}{15} + \dfrac{7}{15} \times \dfrac{8}{15} = \dfrac{112}{225}$

(c) $P(RR) = P(R) \times P(R) = \dfrac{8}{15} \times \dfrac{8}{15} = \dfrac{64}{225}$

Note that $P(WW) + P(RW) + P(RR) = 1$, since the outcomes are exhaustive.

5. A commercial grower rejects his tomatoes for consignment to the wholesaler on three counts: size, shape and colour. By observation over a long period of time he estimates that, on average, out of 1000 tomatoes, he rejects 50 on size, 30 on shape, 20 on colour, 13 on size and shape alone, four on shape and colour alone, one on size and colour alone and one on all three. Calculate the probabilities that a randomly selected tomato: (a) is not rejected; (b) has at least two defects; (c) has not more than one defect.

Call the defective qualities of size, shape and colour A, B and C respectively. For this type of problem the use of a Venn diagram is most helpful (Fig. 1.8). From the diagram we enter 1 in the section common to A, B and C, since there is one reject on all three counts. In the section common to A and B alone we enter 13. In similar fashion we enter 4 in B and C alone and 1 in A and C alone. Then we enter 35,

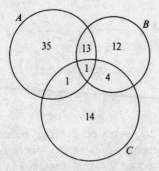

Figure 1.8

12 and 14 in A, B and C alone respectively. These give for A, B and C respective totals of 50, 30 and 20. From the Venn diagram the total number of rejects in 1000 is:

$$35 + 12 + 14 + 13 + 4 + 1 + 1 = 80$$

Therefore the number not rejected must be $1000 - 80 = 920$.

(a) The probability that a randomly selected tomato will not be rejected is $920/1000 = 0.92$.

(b) The number with at least two defects $= 13 + 4 + 1 + 1 = 19$. Therefore the probability that a randomly selected tomato has at least two defects is $19/1000 = 0.019$.

(c) The number of tomatoes with at least two defects and those with not more than one defect represent all possible tomatoes. Therefore the probability of selecting a tomato with not more than one defect $= 1 - 0.019 = 0.981$.

1.4 Mutually exclusive and non mutually exclusive events

Venn diagrams provide an alternative approach to the understanding of the fundamental formulae of the subject.

Non mutually exclusive events

Where outcome A and outcome B of a trial are not mutually exclusive the relationship is illustrated by the Venn diagram shown in Fig. 1.9. The shaded area represents outcomes which are both A and B.

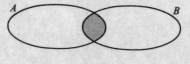

Figure 1.9

Suppose that there are a possible different outcomes A, that there are b possible different outcomes B, and that there are c possible different outcomes A and B. Suppose the total number of possible outcomes of the trial is N. Take $P(A)$ to be the probability that outcome A occurs. Take $P(B)$

to be the probability that outcome B occurs. Take $P(A.B)$ to be the probability that A and B occur together. Take $P(A+B)$ to be the probability that either outcome A alone occurs, or outcome B alone occurs, or both A and B occur. Therefore:

$$P(A+B) = \frac{(a-c)+(b-c)+c}{N}$$

$$= \frac{a+b-c}{N} = \frac{a}{N} + \frac{b}{N} - \frac{c}{N}$$

Now $P(A) = a/N$; $P(B) = b/N$; $P(A.B) = c/N$

Therefore $P(A+B) = P(A) + P(B) - P(A.B)$

Mutually exclusive events

When A and B are mutually exclusive the Venn diagram becomes that of Fig. 1.10. Therefore:

$$c = 0, \text{ i.e. } P(A.B) = 0$$

$$\text{giving } P(A+B) = P(A) + P(B)$$

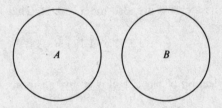

Figure 1.10

Examples

*1. Bag X contains six white and four black balls. Bag Y contains eight white and seven black balls. Bag Z contains 12 white and 19 black balls. One of the three bags, X, Y or Z, is chosen at random. From that bag a ball is chosen at random. Suppose that in the selection of a bag it is equally likely that any one of the three bags is chosen. Take $P(X)$, $P(Y)$ and $P(Z)$ to be the probabilities that bags X, Y and Z are chosen. Then:

$$P(X) = P(Y) = P(Z) = \tfrac{1}{3}$$

The three events are mutually exclusive and exhaustive. They satisfy law *1.10*.

Take $P(W)$ and $P(B)$ to be the probabilities of selecting a white and a black ball respectively. Three different situations arise:

(a) The selection of bag X followed by the choice of a ball from it. Then:

$$P(X) = \frac{1}{3} \text{ and } P(W) = \frac{6}{10} = \frac{3}{5}$$

Therefore, since these events are independent:

$$P(X \text{ and } W) = P(X) \times P(W) = \frac{1}{3} \times \frac{3}{5} = \frac{1}{5}$$

(b) The selection of bag Y followed by the choice of a ball from it. Therefore:

$$P(Y) = \tfrac{1}{3} \text{ and } P(W) = 8/15$$

$$\text{giving } P(Y \text{ and } W) = \frac{1}{3} \times \frac{8}{15}$$

since these events are also independent. That is:

$$P(Y \text{ and } W) = \frac{8}{45}$$

(c) The selection of bag Z followed by the choice of a ball from it. Here:

$$P(Z) = \frac{1}{3} \text{ and } P(W) = \frac{12}{31}$$

$$\text{Therefore } P(Z \text{ and } W) = \frac{1}{3} \times \frac{12}{31} = \frac{4}{31}$$

Consequently the final probability of selecting a white ball is:

$$\frac{1}{5} + \frac{8}{45} + \frac{4}{31}$$

$$= \frac{9+8}{45} + \frac{4}{31} = \frac{17 \times 31 + 4 \times 45}{45 \times 31}$$

$$= \frac{707}{1395}$$

Since the only outcomes can be the choice of a white ball or the choice of a black ball, then:

$$P(W) + P(B) = 1$$

Therefore $P(B) = 1 - \dfrac{707}{1395} = \dfrac{688}{1395}$

2. What is the probability of five tails on five tosses of a fair coin.
   No matter how many tosses of a coin the probability of a tail at each trial is $\frac{1}{2}$. The different trials are independent of each other. Therefore the probability of five tails, i.e. $P(5T)$, is:

   $$\frac{1}{2} \times \frac{1}{2} \times \frac{1}{2} \times \frac{1}{2} \times \frac{1}{2} = \frac{1}{32}$$

3. The probability of throwing a total of seven with two fair dice.
   Seven may be scored as $1+6, 2+5, 3+4, 4+3, 5+2$ or $6+1$. This produces a total of six ways of producing a score of seven. The total number of different possible outcomes $= 6 \times 6 = 36$. Then:

   $$P(7) = \frac{6}{36} = \frac{1}{6}$$

4. What is the probability of not drawing a spade from a standard pack of cards?
   Suppose that $P(S)$ represents the probability of drawing a spade, and that $P(\sim S)$ represents the probability of not drawing a spade. Then $P(S) + P(\sim S) = 1$, since the two outcomes S and $\sim S$ are mutually exclusive and exhaustive. Now:

   $$P(S) = \frac{13}{52} = \frac{1}{4}$$

   Therefore $P(\sim S) = 1 - \dfrac{1}{4} = \dfrac{3}{4}$

*5. In a family of four children, determine the most likely combination of boys and girls.
   Assume it is equally likely that each child may be either a girl or a boy. The different possible combinations are: four boys, $4B$; three boys and one girl, $3B\,1G$; two boys and two girls, $2B\,2G$; one boy and three girls, $1B\,3G$; and four girls, $4G$. The number of possibilities for each child is two. The total number of possible combinations is:

   $$2 \times 2 \times 2 \times 2 = 16$$

   Out of these the number of possibilities for four boys is one.
   The number of possibilities for three boys and one girl is four.

(Imagine the boys spaced *B B B*. Then the girl, *G*, may come in any one of the spaces between two *B*s, before the first *B*, or after the last *B*.) The number of possibilities for two boys and two girls is six. (Space the boys *B B*. Two girls may come before the first *B*, between the two *B*s, or after the last *B*, or two separate *G*s may take up the three spaces in three different ways.)

The number of possibilities for one boy and three girls is four (the same as for three boys and one girl).

The number of possibilities for four girls is one (the same as for four boys). Then:

$$P\,(4B) = P\,(4G) = \frac{1}{16}$$

$$P\,(3B,\,1G) = P\,(1B,\,3G) = \frac{4}{16} = \frac{1}{4}$$

$$P\,(2B,\,2G) = \frac{6}{16} = \frac{3}{8}$$

The most likely combination is two boys and two girls.

Exercise 1.3

Determine the following probabilities.

1. Throwing a four with a fair die.
2. Not throwing a four with a fair die.
3. Throwing less than a four with a fair die.
4. Throwing four or more with a fair die.
5. Throwing more than a four with a fair die.
6. Throwing 12 with two fair dice.
7. Throwing three with two fair dice.
8. Throwing ten with two fair dice.
9. Throwing six with two fair dice.
10. Drawing the ace of spades from a standard pack of cards.
11. Drawing a black card from a standard pack of cards.

In the following questions assume that any card drawn from the pack is replaced before another card is drawn.

12. A black card followed by a red card.
13. A black and a red card in any order.
14. Two cards, one a diamond, the other a club.
15. Two cards each higher than ten (ace is high).
16. Three cards each of the same suit.

17. The six, seven and eight of clubs.

*18. At the end of a shift four mechanics return equal drill bits, indistinguishable from one another, to the store. If the storeman has placed the bits, unmarked, in the same box, what are the probabilities that, on the next shift:
   (a) each man will be given his own drill bit;
   (b) no man will be given his own drill bit;
   (c) two men will be given their own drill bits?

19. Two dice are thrown together. What is the probability that neither is a two or a five?

20. Four coins are tossed together. What are the probabilities that:
   (a) all four coins are head;
   (b) three coins are tail and one is head;
   (c) at least two of the coins are head?

*21. The probability that a man, now aged 50, will be alive x years hence is 5/9. The probability that his wife, aged 48 now, will be alive x years hence is 4/5. Calculate the probabilities that, x years hence:
   (a) only the woman will be alive;
   (b) they will both be alive;
   (c) neither will be alive;
   (d) only one will be alive.

*22. A firm of construction engineers has submitted tenders for three separate metropolitan contracts, A, B and C. The probabilities that the firm will obtain the given contracts are 1/3, 2/11 and 3/5 respectively. Determine the probabilities that it will obtain:
   (a) all three contracts;
   (b) at least two contracts;
   (c) at least one contract;
   (d) no contract.

*23. Two different models, P and Q, of the same design of car, on average, develop signs of rust on the bodywork to differing degrees. The probability that model P develops rust within three years is 1/3. The probability that model Q develops rust within three years is 1/2. Determine the probabilities that, of two cars chosen at random, one of type P and the other of type Q:
   (a) both develop rust within three years;
   (b) neither develops rust within three years;
   (c) at least one develops rust within three years.

*24. A large estate of new houses is being built. The buyers are either local, A, come from between five and ten miles away, B, or come from further afield, C. On average $P(A) = 1/2$, $P(B) = 1/3$ and $P(C) = 1/6$.

The families initially taking up residence have either two members (*D*), three members (*E*), four members (*F*), or more members (*G*), where $P(D) = 1/4$, $P(E) = 1/3$, $P(F) = 7/24$ and $P(G) = 1/8$.

The initial residents leave again either during the first year (*L*), during the second year (*M*), or later (*N*), where $P(L) = 1/20$, $P(M) = 1/10$ and $P(N) = 17/20$.

Calculate the probabilities that a given house will be initially occupied:

(a) by a family with at least three members from not more than ten miles away which will leave again within two years;

(b) by a family with no more than four members from any area which stays more than two years.

*25. An electric circuit is controlled by three separate relays, each of which must function for the circuit to operate. The probabilities that the three relays fail to function are 1/50, 1/40 and 1/60. Determine the probabilities that:

(a) the circuit operates;

(b) at least two of the relays fail.

26. (a) Determine the probability of throwing a total of seven with three dice.

(b) Determine the probability of throwing a five and a two with a pair of dice.

(c) Determine the probability of throwing exactly one two with three dice.

27. *A* and *B* are independent events such that $P(A) = 0.65$ and $P(B) = 0.45$. Determine: (a) $P(A \text{ and } B)$; (b) $P(A \text{ or } B \text{ or both})$.

*28. It is estimated that, on average, 14 per cent of the output of crockery from a pottery are graded seconds. A random sample of four plates is taken. Determine the probabilities that:

(a) all the plates are first grade;

(b) not more than two are seconds;

(c) at least one is a second;

(d) all are seconds

*29. Thirty-five roses in a garden are cut to produce a large, randomly arranged bouquet. Five are red and scented, six are red and not scented, four are yellow and scented, two are yellow and not scented, six are pink and not scented, eight are white and scented and four are white and not scented. A blind man suffering from a heavy cold selects a rose from the bouquet. What are the probabilities:

(a) that it will be scented;

(b) that it will be red;

(c) that it will not be yellow or white?

*30. A private sports club has 300 members. It provides facilities for tennis, croquet and bowls. One hundred and fifty members play croquet, 210 play tennis and 90 play bowls. One hundred and thirty-two play two games only, of whom 30 play just croquet and bowls. A member is selected at random. What are the probabilities that:

(a)   he or she plays tennis only;

(b)   he or she plays all three games?

2

The Binomial Distribution

2.1   The nature of a binomial distribution

Where a particular trial leads to just two different kinds of outcome and no other, then the resulting distribution of probabilities is of a special nature.

Examples

1. The tossing of a fair coin.
    There are two different possible outcomes, H and T (exhaustive events). Then for any particular outcome $P(H)$, i.e. the probability that the result is head, is $\frac{1}{2}$. Also:
    $$P(T) = \tfrac{1}{2} \text{ and } P(H) + P(T) = 1$$

2. The throwing of a three with a fair die.
    There are two different possible outcomes, three or not three (exhaustive events). Then:
    $$P(3) = \frac{1}{6}; \; P(\sim 3) = 1 - \frac{1}{6} = \frac{5}{6}$$
    $$\text{and } P(3) + P(\sim 3) = 1$$

3. The selection of a heart from a standard pack of cards.
    There are two different possible outcomes, a heart or not a heart (exhaustive events). Here:
    $$P(H) = 13/52 = 1/4; \; P(\sim H) = 1 - 1/4 = 3/4;$$
    $$P(H) + P(\sim H) = 1$$

24

4. The probability that a particular manufactured unit will be defective. There are two different possible outcomes: the unit may be defective or not defective (exhaustive events). Suppose that $P(D)$, i.e. the probability that a particular unit chosen at random is defective, is equal to 3/100. Then:

$$P(\sim D) = 1 - 3/100; \; P(D) + P(\sim D) = 1$$

The use of tree diagrams

Suppose that a number of such trials take place in succession and it may be justifiably assumed that the odds at each trial remain unchanged.

For instance, in tossing a fair coin, $P(H)$ remains $\frac{1}{2}$ no matter how many trials take place.

Providing each card drawn from a pack is replaced before another card is drawn at random then the probability of selecting any particular card will always remain the same.

Providing the number of manufactured units from which selections are made is sufficiently large then $P(D)$ will remain the same no matter how many units have been sampled previously, providing, of course, that the sample size is small in comparison with the population size.

The kind of problem we wish to answer is what the probability is that in, say, eight tosses of a fair coin there will be just five heads. One method of approaching simpler problems of this kind is by the employment of tree diagrams.

Examples

1. Determine the probabilities of various possible outcomes of tossing a fair coin twice.
   As shown in Fig. 2.1, the probability of two heads is $\frac{1}{4}$.

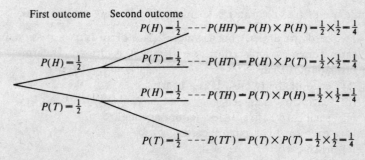

Figure 2.1

The probability of two tails is $\frac{1}{4}$.

The probability of head and tail $= P(HT) + P(TH) = \frac{1}{4} + \frac{1}{4} = \frac{1}{2}$.

Note that at each outcome the sum of the probabilities is 1.

Note also that the sum of the probabilities of the resulting outcomes is 1.

2. Determine the probabilities of throwing two sixes with a pair of dice. As shown in Fig. 2.2, the probability of two sixes is 1/36.

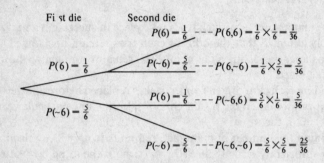

Figure 2.2

The probability of one six $= \dfrac{5}{36} + \dfrac{5}{36} = \dfrac{10}{36}$.

The probability of no sixes $= \dfrac{25}{36}$.

Here again the sum of the probabilities of the individual outcomes is 1 and the sum of the probabilities of the resultant outcomes is also 1.

Exercise 2.1

By the use of tree diagrams determine the following probabilities.

1. Two cards are drawn from a standard pack of cards in succession. Each card drawn is replaced before the next is drawn. Determine the probabilities that:
    (a) each of the two cards is a heart;
    (b) each card is an ace;
    (c) each card is five or less (count ace high);
    (d) each card is a picture card;
    (e) each card is a face card;
    (f) the two cards are consecutive values of the same suit.

2. A bag contains four white and five black balls. A ball is drawn from the bag and then replaced. What is the probability that in two draws a ball of each kind will be selected? When three balls are drawn determine the probability that two will be white. In the latter case determine the probability that at least one black ball will be drawn.

3. In the manufacture of certain kinds of switches it is estimated that 1.5 per cent are defective. In a random sample of three switches calculate the probabilities that:
   (a) no switch is defective;
   (b) at least one will not be defective;
   (c) at least one will be defective.

2.2 Binomial generating functions

Even when the number of outcomes of a trial is as small as three, a tree diagram tends to be complicated. When the number of outcomes is greater the tree diagram method becomes too involved. In such cases we rely on a binomial generating function.

Suppose the number of outcomes of an event is restricted to two. For instance, suppose that one outcome is that A happens (e.g. that a product selected is faulty). Then the only other outcome is that A does not happen (e.g. the product selected is not faulty). Suppose also that $P(A) = p$ no matter how many random selections are made and that $P(\sim A) = q$. Then, since the two outcomes are exclusive and exhaustive:

$$P(A) + P(\sim A) = 1$$
$$\text{that is,} \quad p + q = 1$$
$$\text{or} \quad q = 1 - p$$

The expression $q + pt$ is a binomial expression where t has no significance other than to be associated with the coefficient p, i.e $P(A)$. Then $(q + pt)^n$, where n is a positive integer, is called a binomial generating function.

The coefficient p is often regarded as the probability of success that an event A happens, e.g. that a product selected will be defective. Then the coefficient q is regarded as the probability of failure, i.e. that a product selected will not be defective. Figs 2.3 to 2.5 illustrate the relationship of the binomial generating function to tree diagrams for simple values of n.

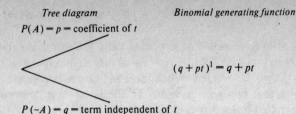

$P(A) = p =$ coefficient of t

$(q + pt)^1 = q + pt$

$P(\sim A) = q =$ term independent of t

Figure 2.3 When $n = 1$

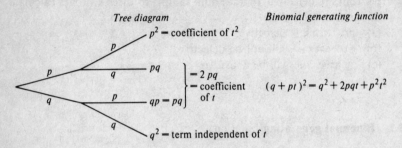

Figure 2.4 When $n = 2$

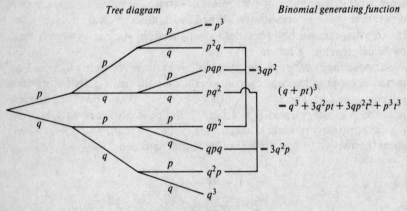

Figure 2.5 When $n = 3$

In other words, in the generating function:

$q^3 =$ term independent of $t =$ probability of three failures of outcome A

$3q^2 p =$ coefficient of $t =$ probability of one success of outcome A

$3qp^2 =$ coefficient of $t^2 =$ probability of two successes of outcome A

$p^3 =$ coefficient of $t^3 =$ probability of three successes of outcome A

By the introduction of the element t we have obtained a valuable indicator of the term which has to be selected to obtain the probability of a given number of successes. *The power of t equals the number of successes.*

Some people regard the introduction of t as unnecessary. They write the binomial generating function as:

$$(q + p)^n$$

In that event it is the power of p which determines the number of successes.

In the examples considered above the number of trials, n, is small and the expansion of the generating function is simple. Where n is not small a special theorem is required to determine the expansion: the binomial theorem.

The binomial theorem

The binomial theorem states that:

$$(q + pt)^n = q^n + \binom{n}{1}q^{n-1}pt + \binom{n}{2}q^{n-2}p^2t^2 + \binom{n}{3}q^{n-3}p^3t^3 + \ldots$$

$$+ \binom{n}{r}q^{n-r}p^rt^r + \ldots + p^nt^n \qquad \qquad 2.1$$

In *2.1* the coefficients are calculated as follows:

$$\binom{n}{1} = \frac{n}{1}; \quad \binom{n}{2} = \frac{n(n-1)}{1.2}; \quad \binom{n}{3} = \frac{n(n-1)(n-2)}{1.2.3};$$

$$\binom{n}{r} = \frac{n(n-1)(n-2)(n-3)\ldots(n-r+1)}{1.2.3\ldots r}$$

Table 2.1 gives the values of the various binomial coefficients for n and r ranging from 1 to 12.

Table 2.1 *Table of binomial coefficients* $\binom{n}{r}$

	0	1	2	3	4	5	6	7	8	9	10	11	12
1	1	1											
2	1	2	1										
3	1	3	3	1									
4	1	4	6	4	1								
5	1	5	10	10	5	1							
6	1	6	15	20	15	6	1						
7	1	7	21	35	35	21	7	1					
8	1	8	28	56	70	56	28	8	1				
9	1	9	36	84	126	126	84	36	9	1			
10	1	10	45	120	210	252	210	120	45	10	1		
11	1	11	55	165	330	462	462	330	165	55	11	1	
12	1	12	66	220	495	792	924	792	495	220	66	12	1

(n labels the rows)

Examples

1. The generating function for selecting a spade from a standard pack of cards when the number of draws is four, and when each card drawn is replaced before another one is drawn, is:

$$\left(\frac{3}{4}+\frac{1}{4}t\right)^4$$

2. The generating function for throwing a six with a fair die when the number of throws is n is.

$$\left(\frac{5}{6}+\frac{1}{6}t\right)^n$$

3. A fair coin is tossed seven times. Determine the probabilities that the following results are obtained:

    (a) four heads and three tails;
    (b) two heads and five tails;
    (c) at least three heads.

The probability generating function is:

$$\left(\frac{1}{2}+\frac{1}{2}t\right)^7$$

$$= \left(\frac{1}{2}\right)^7 + \binom{7}{1}\left(\frac{1}{2}\right)^6\left(\frac{1}{2}t\right) + \binom{7}{2}\left(\frac{1}{2}\right)^5\left(\frac{1}{2}t\right)^2 + \binom{7}{3}\left(\frac{1}{2}\right)^4\left(\frac{1}{2}t\right)^3 +$$

$$\binom{7}{4}\left(\frac{1}{2}\right)^3\left(\frac{1}{2}t\right)^4 + \binom{7}{5}\left(\frac{1}{2}\right)^2\left(\frac{1}{2}t\right)^5 + \binom{7}{6}\left(\frac{1}{2}\right)\left(\frac{1}{2}t\right)^6 + \left(\frac{1}{2}t\right)^7$$

$$= \frac{1}{128} + 7\cdot\frac{1}{128}t + \frac{7.6}{1.2}\cdot\frac{1}{128}t^2 + \frac{7.6.5}{1.2.3}\cdot\frac{1}{128}t^3 + \frac{7.6.5.4}{1.2.3.4}\cdot\frac{1}{128}t^4$$

$$+ \frac{7.6.5.4.3}{1.2.3.4.5}\cdot\frac{1}{128}t^5 + \frac{7.6.5.4.3.2}{1.2.3.4.5.6}\cdot\frac{1}{128}t^6 + \frac{1}{128}t^7$$

$$= \frac{1}{128} + \frac{7}{128}t + \frac{21}{128}t^2 + \frac{35}{128}t^3 + \frac{35}{128}t^4 + \frac{21}{128}t^5 + \frac{7}{128}t^6 + \frac{1}{128}t^7 \quad \text{(i)}$$

Alternatively the expansion (i) may be obtained as follows:

$$\left(\frac{1}{2}+\frac{1}{2}t\right)^7 = \left(\frac{1}{2}\right)^7(1+t)^7 = \frac{1}{128}(1+t)^7$$

$(1+t)^7$ may be expanded using Table 2.1. This produces:

$$\frac{1}{128}(1+7t+21t^2+35t^3+35t^4+21t^5+7t^6+t^7)$$

Now suppose that $P(H) = p$ is success and $P(T) = q$ is failure, where $p = q = \frac{1}{2}$.

(a) $P(4H, 3T)$ = coefficient of t^4 in (i) = $\frac{35}{128}$
(b) $P(2H, 5T)$ = coefficient of t^2 in (i) = $\frac{21}{128}$
(c) The probability of at least three heads is:

$$P(3H, 4T) + P(4H, 3T) + P(5H, 2T) + P(6H, T) + P(7H)$$

= sum of coefficients of t^3, t^4, t^5, t^6, t^7 in (i)

$$= \frac{35}{128} + \frac{35}{128} + \frac{21}{128} + \frac{7}{128} + \frac{1}{128}$$

$$= \frac{99}{128}$$

4. Suppose the probability that a particular brand of lamp is defective is 2 per cent. A sample of ten such lamps is chosen at random. Determine the probabilities that the sample contains:

(a) exactly one defective lamp;
(b) exactly two defective lamps;
(c) not more than five defective lamps;
(d) at least three defective lamps.

The probability of success, p, equals the probability that a lamp is defective, i.e. 2 per cent = 2/100 = 1/50.
The probability of failure, $q = 1 - p = 49/50$.
The generating function is $\left(\dfrac{49}{50} + \dfrac{1}{50}t \right)^{10} \equiv g(t)$.

(a) The probability of exactly one defective lamp is:

$$P(1) = \text{coefficient of } t \text{ in the expansion of } g(t)$$

$$= \binom{10}{1}\left(\frac{49}{50}\right)^9\left(\frac{1}{50}\right) = \frac{1}{5}\left(\frac{49}{50}\right)^9$$

$$= \frac{1}{5}(0.98)^9 = 0.1667496 \text{ by calculator}$$

(b) The probability of exactly two defective lamps is $P(2)$, which is:

$$\text{coefficient of } t^2 \text{ in the expansion of } g(t)$$

$$= \binom{10}{2}\left(\frac{49}{50}\right)^8\left(\frac{1}{50}\right)^2$$

$$= \frac{10.9}{1.2}(0.98)^8(0.02)^2$$

$$= 45(0.98)^8(0.02)^2 = 0.0153137 \text{ by calculator}$$

*(c)   The probability of not more than five defective lamps is:

$$P(0) + P(1) + P(2) + P(3) + P(4) + P(5) \text{ where}$$
$$P(0) = (0.98)^{10} = 0.8170728$$
$$P(1) = 0.1667496, \text{ from above}$$
$$P(2) = 0.0153137, \text{ from above}$$
$$P(3) = \binom{10}{3}(0.98)^7(0.02)^3$$
$$= 120(0.98)^7(0.02)^3$$
$$= 0.0008334$$
$$P(4) = \binom{10}{4}(0.98)^6(0.02)^4$$
$$= 210(0.98)^6(0.02)^4$$
$$= 0.0000298$$
$$P(5) = \binom{10}{5}(0.98)^5(0.02)^5$$
$$= 252(0.98)^5(0.02)^5$$
$$= 0.0000007$$

The probability required is obtained by adding the individual probabilities determined above. It equals 1. In other words, it is reasonably certain that not more than five defective lamps will be selected.

*(d)   The probability of at least three defective lamps is:

$$1 - P(0) - P(1) - P(2)$$
$$= 1 - 0.8170728 - 0.1667496 - 0.0153137$$
$$= 0.0008639$$

Therefore the probability that there will be at least three defective lamps among the sample is slight.

Exercise 2.2

Write down generating functions for the following.

1.   The probability of throwing at most a two with a fair die for four throws.

2. The probability of throwing at least a two with a fair die in five throws.
3. The probability of drawing an ace from a standard pack of cards in four draws when every card drawn is replaced before another card is drawn.
4. A fair coin is tossed four times. Determine the probabilities of the following outcomes:
    (a) three heads;
    (b) two heads;
    (c) at least one head.
*5. A fair die is thrown six times. Determine the probabilities of the following outcomes:
    (a) exactly three sixes;
    (b) at least one six;
    (c) no throw is less than three;
    (d) all throws are greater than two;
    (e) there are exactly four twos.
*6. A card is drawn from a pack eight times and each time replaced before another one is drawn. Calculate the probabilities of the following outcomes:
    (a) four tens;
    (b) three aces;
    (c) eight picture cards;
    (d) no card is higher than a six (count ace low).
*7. On the 100-metres range a particular marksman scores a bull on average three times in four shots. For that range and for that marksman, determine the probabilities of the following outcomes:
    (a) exactly seven bulls in eight shots;
    (b) not less than five bulls in eight shots;
    (c) exactly ten bulls in 11 shots;
    (d) not less than three bulls in ten shots.
*8. Similar packets of seeds of the same variety of lobelia plant produce two colours only. On average 3/5 of the plants have dark blue flowers; the rest have pale blue flowers. Determine the probabilities of the following outcomes for randomly selected plants:
    (a) out of 20 plants exactly 16 will be dark blue;
    (b) out of 20 plants not more than eight will be pale blue;
    (c) out of 20 plants not more than eight will be dark blue.
*9. On average a firm producing electric typewriters finds that 5 per cent of them develop faults within 12 months of sale. A retailer orders 20 typewriters. Determine:
    (a) the probability that exactly three of them will prove faulty within 12 months;

(b) the probability that none of them will be faulty within 12 months of sale;

(c) the probability that not more than four of them will be faulty within the twelve-month period.

2.3 Mean and standard deviation of a binomial distribution

For the binomial distribution defined by the generating function

$$(q + pt)^n$$

in which p represents the probability of success of any particular outcome and $q = 1 - p$, then the mean of that distribution is:

$$np \qquad\qquad 2.2$$

and the standard deviation is:

$$\sqrt{npq} \qquad\qquad 2.3$$

Therefore the variance, which is the square of the standard deviation, is:

$$npq \qquad\qquad 2.4$$

There are varying symbols in use for these measures. The ones used here are: μ for the mean; σ for standard deviation; and σ^2 for variance.

* *To verify formula 2.2 when n = 3*

When $n = 3$ the number of possible successes may be 0, 1, 2 or 3. The probabilities of these successes are q^3, $3q^2 p$, $3qp^2$ and p^3 respectively. Then the mean, μ, is given by:

$$\mu(q^3 + 3q^2 p + 3qp^2 + p^3)$$
$$= 0 \times q^3 + 1 \times 3q^2 p + 2 \times 3qp^2 + 3 \times p^3$$
$$= 3p(q^2 + 2qp + p^2)$$
$$= 3p(q + p)^2$$

i.e. $\mu(q + p)^3 = 3p(q + p)^2$

i.e. $\mu = 3p$, since $q + p = 1$

** *To prove formula 2.2 for general n*

The number of possible successes may be 0, 1, 2, 3, . . . r, . . . n. The probabilities of these outcomes are:

$$q^n, \binom{n}{1}q^{n-1}p, \binom{n}{2}q^{n-2}p^2, \ldots \binom{n}{r}q^{n-r}p^r, \ldots p^n$$

Suppose the mean is μ. Then:

$$\mu\left[q^n + \binom{n}{1}q^{n-1}p + \binom{n}{2}q^{n-2}p^2 + \ldots + \binom{n}{r}q^{n-r}p^r + \ldots p^n \right]$$

$$= 0 \times q^n + 1 \times \binom{n}{1}q^{n-1}p + 2 \times \binom{n}{2}q^{n-2}p^2 + 3 \times \binom{n}{3}q^{n-3}p^3$$

$$+ \ldots + r \times \binom{n}{r}q^{n-r}p^r + \ldots + np^n$$

$$= \binom{n}{1}q^{n-1}p + \binom{n}{1}q^{n-2}p^2 + \binom{n}{2}q^{n-3}p^3 + \ldots$$

$$+ \binom{n}{r-1}q^{n-r}p^r + \ldots + np^n$$

$$= np\left[q^{n-1} + \binom{n-1}{1}q^{n-2}p + \binom{n-1}{2}q^{n-3}p^2 + \ldots \right.$$

$$\left. + \binom{n-1}{r-1}q^{n-r}p^{r-1} + \ldots + p^{n-1} \right]$$

i.e. $\mu(q+p)^n = np(q+p)^{n-1}$

or $\mu = np$

* *To verify formula 2.3 when $n = 3$*

The individual deviations from the mean, $3p$, are $(0-3p)$, $(1-3p)$, $(2-3p)$, $(3-3p)$, i.e. $-3p$, $(1-3p)$, $(3q-1)$, $3q$, using $p+q = 1$. Then the variance, σ^2, is:

$$(-3p)^2q^3 + (1-3p)^2 3q^2p + (3q-1)^2 3qp^2 + (3q)^2 \cdot p^3$$

$$= 9p^2q^3 + 9q^2p^3 + 3pq.[q(1-3p)^2 + p(3q-1)^2]$$

$$= 9p^2q^2(p+q) + 3pq.[(q+p) - 12pq + 9pq(p+q)]$$

$$= 9p^2q^2 + 3pq.[1 - 3pq]$$

$$= 9p^2q^2 + 3pq - 9p^2q^2$$

$$= 3pq$$

The standard deviation is $\sqrt{3pq}$.

** *To prove formula 2.3*

One standard formula for the variance of a frequency distribution is:

$$\frac{1}{N} \sum_{r=0}^{n} f_r x_r^2 - \bar{x}^2$$

where $\bar{x}$ is the mean. In applying this formula to the binomial distribution x_r takes on the values of the numbers of possible successes, i.e. 0, 1, 2, 3, ... n, so that $x_r = r$. The values of f_r are the probabilities of the corresponding successes, i.e.

$$f_r = \binom{n}{r} q^{n-r} p^r$$

And $N = 1$, since $N =$ the sum of all probabilities, which is:

$$(q+p)^n = 1^n = 1$$

Then the variance is:

$$\sum_{r=0}^{n} \binom{n}{r} q^{n-r} p^r . r^2 - n^2 p^2$$

$$= \sum_{r=1}^{n} \binom{n}{r} q^{n-r} p^r r^2 - n^2 p^2$$

$$= np \sum_{r=1}^{n} \binom{n-1}{r-1} q^{n-r} p^{r-1} r - n^2 p^2$$

$$= np \sum_{r=1}^{n} \left\{ \binom{n-1}{r-1} q^{n-r} p^{r-1} [(r-1)+1] \right\} - n^2 p^2$$

$$= np \sum_{r=1}^{n} \binom{n-1}{r-1} q^{n-r} p^{r-1} (r-1) + np \sum_{r=1}^{n} \binom{n-1}{r-1} q^{n-r} p^{r-1} .1 - n^2 p^2$$

$$= np \sum_{r=2}^{n} \binom{n-1}{r-1} q^{n-r} p^{r-1} (r-1) + np (q+p)^{n-1} - n^2 p^2$$

$$= (n-1)np \sum_{r=2}^{n} \binom{n-2}{r-2} q^{n-r} p^{r-1} + np - n^2 p^2$$

$$= (n-1)np^2 + np - n^2 p^2$$

$$= n^2 p^2 - np^2 + np - n^2 p^2$$

$$= np - np^2 = np(1-p) = npq$$

The standard deviation is $\sqrt{npq}$.

** *Example*

When n is constant determine the value of p to produce the maximum standard deviation.

Here represent the standard deviation by D. Then:

$$D^2 = np(1-p), \text{ by formula}$$
$$= np - np^2 \qquad\qquad 2.3$$

Differentiate w.r.t. p. The LHS is a function of a function of p.

$$2D \cdot \frac{dD}{dp} = n - 2np$$

$$\frac{dD}{dp} = \frac{n(1-2p)}{2D} = 0 \text{ for stationary values}$$

that is, $n(1-2p) = 0$ for stationary values

Therefore $1 - 2p = 0$

i.e. $p = \frac{1}{2}$

When $p = \frac{1}{2}$ $D \neq 0$. To determine the nature of the stationary value:

when $p < \frac{1}{2}$ $\quad dD/dp > 0$
when $p > \frac{1}{2}$ $\quad dD/dp < 0$

Therefore, when $p = \frac{1}{2}$, D takes a maximum value.

2.4 Bar chart representation of a binomial distribution

The representation of a binomial distribution may be expressed satisfactorily in bar chart form.

Examples

1. Represent the distribution $(\frac{1}{2} + \frac{1}{2})^{12}$ in bar chart form.
   Here the probability of success for any outcome is $\frac{1}{2}$. In 12 trials the numbers of possible successes are 0, 1, 2, 3, 4, . . . 12. The corresponding probabilities are given below:

$$P(0) = \left(\frac{1}{2}\right)^{12} \qquad\qquad = 0.0002441$$

$$P(1) = \binom{12}{1}\left(\frac{1}{2}\right)^{11}\left(\frac{1}{2}\right) = 12\left(\frac{1}{2}\right)^{12} \qquad = 0.0029297$$

$$P(2) = \binom{12}{2}\left(\frac{1}{2}\right)^{12} = \frac{12.11}{1.2}\left(\frac{1}{2}\right)^{12} \qquad = 0.0161133$$

$$P(3) = \binom{12}{3}\left(\frac{1}{2}\right)^{12} = \frac{12!}{9!\,3!}\left(\frac{1}{2}\right)^{12} \qquad = 0.0537109$$

$$P(4) = \binom{12}{4}\left(\frac{1}{2}\right)^{12} = \frac{12!}{8!\,4!}\left(\frac{1}{2}\right)^{12} \qquad = 0.1208496$$

$$P(5) = \binom{12}{5}\left(\frac{1}{2}\right)^{12} = \frac{12!}{7!\,5!}\left(\frac{1}{2}\right)^{12} \qquad = 0.1933594$$

$$P(6) = \binom{12}{6}\left(\frac{1}{2}\right)^{12} = \frac{12!}{6!\,6!}\left(\frac{1}{2}\right)^{12} \qquad = 0.2255859$$

$$P(7) = P(5) \qquad\qquad = 0.1933594$$
$$P(8) = P(4) \qquad\qquad = 0.1208496$$
$$P(9) = P(3) \qquad\qquad = 0.0537109$$
$$P(10) = P(2) \qquad\qquad = 0.0161133$$
$$P(11) = P(1) \qquad\qquad = 0.0029297$$
$$P(12) = P(0) \qquad\qquad = 0.0002441$$

The mean $= np = 12 \times \frac{1}{2} = 6$

The standard deviation $= \sqrt{npq} = \sqrt{12 \times \frac{1}{2} \times \frac{1}{2}} = \sqrt{3}$

This distribution is represented in Fig. 2.6.

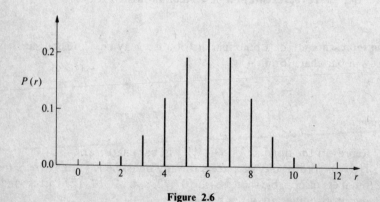

Figure 2.6

2. Represent the distribution $(0.8 + 0.2)^{10}$ in diagrammatic form.

$$P(0) = (0.8)^{10} \qquad\qquad = 0.1\bar{7}03742$$

$$P(1) = 10(0.8)^9(0.2) \qquad = 0.2684355$$

$$P(2) = \frac{10!}{8!\,2!}(0.8)^8(0.2)^2 \quad = 0.3019899$$

$$P(3) = \frac{10!}{7!\,3!}(0.8)^7(0.2)^3 \cdot \;\; = 0.2013266$$

$$P(4) = \frac{10!}{6!\,4!}(0.8)^6(0.2)^4 \quad = 0.0880804$$

$$P(5) = \frac{10!}{5!\,5!}(0.8)^5(0.2)^5 \quad = 0.0264241$$

$$P(6) = \frac{10!}{4!\,6!}(0.8)^4(0.2)^6 \quad = 0.005505$$

$$P(7) = \frac{10!}{3!\,7!}(0.8)^3(0.2)^7 \quad = 0.0007864$$

$$P(8) = \frac{10!}{2!\,8!}(0.8)^2(0.2)^8 \quad = 0.0000737$$

$$P(9) = \frac{10!}{1!\,9!}(0.8)(0.2)^9 \quad = 0.0000041$$

$$P(10) = (0.2)^{10} \qquad\qquad = 0.0000001$$

Mean $= 10 \times 0.2 = 2$

Standard deviation $= \sqrt{10 \times 0.8 \times 0.2} = \sqrt{1.6} = 1.2649111 \approx 1.265$

This distribution is represented in Fig. 2.7.

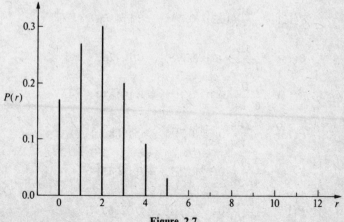

Figure 2.7

3. Represent the distribution $(0.3 + 0.7)^{20}$ in diagrammatic form.

$$P(0) = (0.3)^{20} = 3.4867844 \times 10^{-11}$$

$$P(1) = 20(0.3)^{19}(0.7) = 1.6271661 \times 10^{-9}$$

$$P(2) = \frac{20!}{18!\,2!}(0.3)^{18}(0.7)^2 = 3.606884 \times 10^{-8}$$

$$P(3) = \frac{20!}{17!\,3!}(0.3)^{17}(0.7)^3 = 0.0000005$$

$$P(4) = \frac{20!}{16!\,4!}(0.3)^{16}(0.7)^4 = 0.000005$$

$$P(5) = \frac{20!}{15!\,5!}(0.3)^{15}(0.7)^5 = 0.0000374$$

$$P(6) = \frac{20!}{14!\,6!}(0.3)^{14}(0.7)^6 = 0.0002181$$

$$P(7) = \frac{20!}{13!\,7!}(0.3)^{13}(0.7)^7 = 0.0005089$$

$$P(8) = \frac{20!}{12!\,8!}(0.3)^{12}(0.7)^8 = 0.0038593$$

$$P(9) = \frac{20!}{11!\,9!}(0.3)^{11}(0.7)^9 = 0.0120067$$

$$P(10) = \frac{20!}{10!\,10!}(0.3)^{10}(0.7)^{10} = 0.0308171$$

$$P(11) = \frac{20!}{9!\,11!}(0.3)^9(0.7)^{11} = 0.0653696$$

$$P(12) = \frac{20!}{8!\,12!}(0.3)^8(0.7)^{12} = 0.1143967$$

$$P(13) = \frac{20!}{7!\,13!}(0.3)^7(0.7)^{13} = 0.164262$$

$$P(14) = \frac{20!}{6!\,14!}(0.3)^6(0.7)^{14} = 0.191639$$

$$P(15) = \frac{20!}{5!\,15!}(0.3)^5(0.7)^{15} = 0.1788631$$

$$P(16) = \frac{20!}{4!\,16!}(0.3)^4(0.7)^{16} = 0.130421$$

$$P(17) = \frac{20!}{3!\,17!}(0.3)^3(0.7)^{17} = 0.0716037$$

$$P(18) = = \frac{20!}{2!\,18!}(0.3)^2(0.7)^{18} = 0.0278459$$

$$P(19) = \frac{20!}{1!\,19!}(0.3)(0.7)^{19} = 0.0068393$$

$$P(20) = (0.7)^{20} = 0.0007979$$

Mean $= 20 \times 0.7 = 14$

Standard deviation $= \sqrt{20 \times 0.7 \times 0.3} = 2.0493902 \approx 2.049$

Fig. 2.8 is a diagrammatic representation of this distribution.

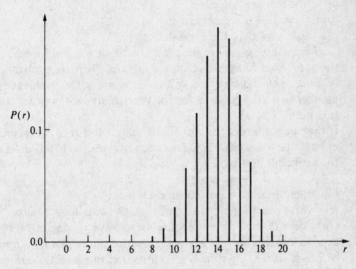

Figure 2.8

Exercise 2.3

Represent the following binomial distributions in diagrammatic form. In each case determine the mean and the standard deviation.

1. $(0.6 + 0.4)^4$
2. $(0.6 + 0.4)^{10}$
3. $(0.4 + 0.6)^{12}$
4. $(0.7 + 0.3)^{10}$
5. $(0.2 + 0.8)^{10}$
6. $(\frac{1}{2} + \frac{1}{2})^{10}$
7. $(\frac{3}{4} + \frac{1}{4})^6$

8. $(\frac{1}{4} + \frac{3}{4})^{12}$

9. $(0.9 + 0.1)^{10}$

10. $(0.99 + 0.01)^{10}$

Exercise 2.4

**1. On average 8 per cent of the electric switches produced by a particular unit in a factory are faulty. A random sample of ten switches is taken. Determine the probabilities that:

    (a)   just two are defective;

    (b)   at most two are defective;

    (c)   not more than three are defective.

**2. Steel girders which are to function as struts are tested by subjecting them to a huge compressive impulsive force. There is a probability of 0.75 that each and every test will distort a girder. Determine the least number of tests necessary for the probability that a girder will be fractured to be 0.95.

**3. In the game of bridge the probability that a hand of 13 cards contains at least one ace is 0.6962. In ten randomly selected hands determine the probabilities that:

    (a)   no ace is ever present;

    (b)   there is at least one ace in each hand;

    (c)   there are no more than three hands containing no ace.

**4. On a certain long-distance route an average of one in every five express coaches arrives at its destination more than 15 minutes before schedule. Eight days are selected at random. Determine the probabilities that:

    (a)   all eight coaches arrive more than 15 minutes before time;

    (b)   only two coaches arrive more than 15 minutes ahead of schedule;

    (c)   at least five coaches arrive more than 15 minutes ahead of schedule.

3

The Poisson Distribution

Whenever the binomial distribution is characterized by a large value of n and a value of p which is small, the calculation of the various probabilities becomes tedious. The examples at the end of Chapter 2 bear out that fact. Consequently it is more convenient to use the Poisson distribution, which is applicable in such circumstances providing np remains constant.

3.1 The behaviour of the binomial distribution of large samples of a population when all samples have the same mean

Consider a very large population of eggs, some of which have double yolks. Suppose random samples of that population, of sizes 100, 1000 and 10 000, are taken, in which there are 2 per cent, 0.2 per cent and 0.02 per cent double-yolked eggs, i.e. $p = 0.02$, 0.002 and 0.0002 respectively. The mean, in each case, is np. The values of np are 2, 2 and 2.

In other words, although the percentages (or proportions) of double-yolked eggs in the various samples differ from each other, the means are the same. The binomial distributions are:

(a) $(0.98 + 0.02)^{100}$
(b) $(0.998 + 0.002)^{1000}$
(c) $(0.9998 + 0.0002)^{10\,000}$

For (a):

$$P(0) = 0.1326$$
$$P(1) = 0.2707$$
$$P(2) = 0.2734$$
$$P(3) = 0.1823$$
$$P(4) = 0.0902$$
$$P(5) = 0.0353$$

For (b):

$$P(0) = 0.1351$$
$$P(1) = 0.2707$$
$$P(2) = 0.2704$$
$$P(3) = 0.1806$$
$$P(4) = 0.0902$$
$$P(5) = 0.0360$$

For (c):

$$P(0) = 0.1353$$
$$P(1) = 0.2707$$
$$P(2) = 0.2707$$
$$P(3) = 0.1805$$
$$P(4) = 0.0902$$
$$P(5) = 0.0361$$

There is close agreement between the three different distributions.

Before we determine the nature of the Poisson distribution we shall need to look at an alternative definition of e.

3.2 Alternative definition of e

In *Analytical Mathematics 2* e^x is defined as that function whose derivative is itself. The quantity e may also be defined as:

$$\text{Lt}_{n \to \infty} \left(1 + \frac{1}{n}\right)^n \qquad 3.1$$

Therefore:

$$\text{Lt}_{n \to \infty} \left(1 + \frac{x}{n}\right)^n = e^x \qquad 3.2$$

We may verify that *3.1* is a reasonable definition by means of a few simple calculations.

Examples

1. The value of $\left(1 + \dfrac{1}{10}\right)^{10} = 1.1^{10} = 2.5937425$

2. The value of $\left(1 + \dfrac{1}{100}\right)^{100} = 1.01^{100} = 2.7048138$

3. The value of $\left(1 + \dfrac{1}{1000}\right)^{1000} = 1.001^{1000} = 2.716924$

4. The value of $\left(1 + \dfrac{1}{10^6}\right)^{10^6} = 2.7182818$

The calculator gives the value of e as 2.7182818. Similar calculations verify that formula *3.2* is reasonable.

Examples

1. The value of $\left(1 + \dfrac{3}{10^6}\right)^{10^6} = 20.085537$

   The value of $e^3 = 20.085537$

2. The value of $\left(1 + \dfrac{17.8}{10^6}\right)^{10^6} = 53\,750\,042$

   The value of $e^{17.8} = 53\,757\,836$

3. The value of $\left(1 - \dfrac{14.7}{10^6}\right)^{10^6} = 0.0000004$

   The value of $e^{-14.7} = 0.0000004$

Therefore, by *3.2*:

$$\underset{n \to \infty}{\text{Lt}} \left[1 + \frac{\mu(t-1)}{n}\right]^n = e^{\mu(t-1)}$$

$$= e^{-\mu} \cdot e^{\mu t} \qquad\qquad 3.3$$

** 3.3 To obtain the Poisson distribution

When the probability of success of an outcome is p and when the number of trials is n the binomial generating function is:

$$g(t) \equiv (q + pt)^n$$
$$\equiv [(1-p) + pt]^n, \text{ since } q + p = 1$$
$$\equiv [1 + p(t-1)]^n$$

Suppose $np = \text{constant} = \mu$, $p = \mu/n$. Then:

$$g(t) \equiv \left[1 + \frac{\mu}{n}(t-1)\right]^n$$

As n increases indefinitely:

$$\underset{n \to \infty}{\text{Lt}} \; [g(t)] \equiv \underset{n \to \infty}{\text{Lt}} \left[1 + \frac{\mu(t-1)}{n}\right]^n$$
$$= e^{-\mu} \cdot e^{\mu t} \quad \text{(by 3.3)}$$

However, by the exponential expansion:

$$e^{\mu t} = 1 + \mu t + \frac{1}{2!}(\mu t)^2 + \frac{1}{3!}(\mu t)^3 + \ldots \frac{1}{r!}(\mu t)^r + \ldots$$

So
$$g(t) = e^{-\mu} + e^{-\mu} \cdot \mu t + e^{-\mu} \cdot \frac{1}{2!}(\mu t)^2 + \ldots + e^{-\mu} \cdot \frac{1}{r!}(\mu t)^r + \ldots \quad 3.4$$

The series *3.4* represents the Poisson distribution for which the mean is μ. A fundamental characteristic of this distribution is that p does not explicitly occur in the series. The means of the distribution must be known or be determinable before the series may be used. The coefficients of the powers in the individual terms indicate the probabilities of associated numbers of successful outcomes of the event involved.

Suppose that series *3.4* represents the Poisson distribution of a large population of which the mean is μ. Suppose also that $P(0)$ represents the probability that, in a large sample, the number of successful outcomes is 0, $P(1)$ represents the probability of one successful outcome, $P(2)$ represents the probability of two successful outcomes, and, in general, $P(r)$ represents the probability of r successful outcomes. Then:

$$P(0) = e^{-\mu}$$
$$P(1) = e^{-\mu} \cdot \mu$$
$$P(2) = e^{-\mu} \cdot \frac{1}{2!}\mu^2$$
$$P(3) = e^{-\mu} \cdot \frac{1}{3!}\mu^3$$
$$P(r) = e^{-\mu} \cdot \frac{1}{r!}\mu^r$$

Examples

1. Determine the values of $P(r)$ for the Poisson distribution with mean 2 for values of r from 0 to 5 inclusive.
   By the formulae above:

   $$P(0) = e^{-2} \qquad\qquad = 0.1353353 \approx 0.1353$$

   $$P(1) = e^{-2}.2 \qquad\qquad = 0.2706706 \approx 0.2707$$

   $$P(2) = e^{-2}.\frac{1}{2!}.2^2 \qquad = 0.2706706 \approx 0.2707$$

   $$P(3) = e^{-2}.\frac{1}{3!}2^3 \qquad = 0.180447 \approx 0.1804$$

   $$P(4) = e^{-2}.\frac{1}{4!}2^4 \qquad = 0.0902235 \approx 0.0902$$

   $$P(5) = e^{-2}.\frac{1}{5!}2^5 \qquad = 0.0360894 \approx 0.0361$$

   A comparison between these probabilities and those of the binomial distributions in section 3.1 illustrates how close the resemblance is.

2. On average, over a period of many years in a large region of Britain, there are five major Premium-bond prize-winners in a year. Determine the probabilities that, in a year chosen at random, there will be, in the region:
   (a) no major prize-winners;
   (b) three major prize-winners;
   (c) not more than six major prize-winners;
   (d) at least four major prize-winners.
   In this case $\mu = 5$. Therefore:

   $$P(0) = e^{-5} \qquad\qquad = 0.0067379$$

   $$P(1) = e^{-5}.5 \qquad\qquad = 0.0336897$$

   $$P(2) = e^{-5}.\frac{1}{2!}5^2 \qquad = 0.0842243$$

   $$P(3) = e^{-5}.\frac{1}{3!}5^3 \qquad = 0.1403739$$

   $$P(4) = e^{-5}.\frac{1}{4!}5^4 \qquad = 0.1754674$$

   $$P(5) = e^{-5}.\frac{1}{5!}5^5 \qquad = 0.1754674$$

   $$P(6) = e^{-5}.\frac{1}{6!}5^6 \qquad = 0.1462228$$

(a)   The probability is $P(0) = 0.0067379$
(b)   The probability is $P(3) = 0.1403739$
(c)   The probability is:

$$P(0) + P(1) + P(2) + P(3) + P(4) + P(5) + P(6)$$
$$= 0.7621834$$

(d)   The probability is:

$$1 - P(0) - P(1) - P(2) - P(3)$$
$$= 0.7349742$$

3.   In a very large city, on average during a year, 2 per cent of the insured householders make a claim on the insurance companies for some form of accidental damage. One hundred and twenty insured householders in the city are selected at random. Determine the probabilities that in that sample in one year:

(a)   no householder makes a claim;
(b)   exactly three make a claim;
(c)   not less than four make a claim;
(d)   not more than two make a claim.

In this problem p, the proportion of householders claiming is given:

$$p = 2 \text{ per cent} = 2/100 = 1/50$$

The sample number, n, is 120.

Therefore, since $\mu = np$, $\mu = 120 \times \dfrac{1}{50} = 2.4$. Therefore:

$$P(0) = e^{-2.4} \qquad\qquad\qquad = 0.90718$$
$$P(1) = e^{-2.4} \times 2.4 \qquad\qquad = 0.2177231$$
$$P(2) = e^{-2.4} \times \frac{1}{2!}(2.4)^2 \quad\; = 0.2612677$$
$$P(3) = e^{-2.4} \times \frac{1}{3!}(2.4)^3 \quad\; = 0.2090142$$
$$P(4) = e^{-2.4} \times \frac{1}{4!}(2.4)^4 \quad\; = 0.1254085$$

(a)   The probability is $P(0) = 0.090718$.
(b)   The probability is $P(3) = 0.2090142$.
(c)   The probability is $1 - P(0) - P(1) - P(2) - P(3) = 0.2212771$.
(d)   The probability is $P(0) + P(1) + P(2) = 0.5697087$.

Exercise 3.1

For distributions of the Poisson model calculate the following probabilities for the given values of μ.

1. $\mu = \frac{1}{2}$; $P(0)$, $P(1)$, $P(2)$, $P(3)$. Compare these values with the corresponding probability values using the binomial distribution when $n = 100$.

2. $\mu = \frac{1}{4}$; $P(0)$, $P(1)$, $P(2)$, $P(3)$, $P(4)$. Compare these values with the corresponding probability values using the binomial distribution when $n = 1000$.

3. $\mu = 1/10$; $P(0)$, $P(1)$, $P(2)$, $P(3)$. Compare these values with the corresponding probability values using the binomial distribution when $n = 500$.

4. $\mu = 1/100$; $P(0)$, $P(1)$, $P(2)$, $P(3)$. Then calculate $P(0) + P(1) + P(2) + P(3)$.

5. Suppose that on average one house in every 100 in a particular area has a fire in any year. There is a total of 1000 houses in that area. Calculate the probabilities that in this area:
   (a) no house has a fire during one year;
   (b) one house has a fire during one year;
   (c) two houses have fires during one year;
   (d) ten houses have fires during one year;
   *(e) not more than 11 houses have fires during one year.

6. A book of 350 pages has a total of 35 misprints scattered at random throughout the book. Determine the probability that a page of the book chosen at random has:
   (a) no misprints;
   (b) two misprints;
   (c) at least three misprints.

7. On average 1 per cent of the model electric engines produced by a firm are defective. They are dispatched to wholesalers in batches of 100. Determine the probabilities that in a particular batch there are:
   (a) no defective engines;
   (b) no more than one defective engine;
   (c) at least two defective engines.

8. A fruit farm sends large quantities of choice quality apples to a packing station. On average one apple in every 200 is bad. The apples are packed in boxes of 50. Determine the probabilities that in a selected box there will be:
   (a) no bad apples;
   (b) at least three bad apples;
   (c) not more than two bad apples.

3.4 Events which meet the Poisson requirements

Certain circumstances must be operative in order for the Poisson distribution to give a reasonable fit to the actual probabilities:
1. The population must be large.
2. The events occur singly along the time scale or other axis when time is not the independent variable.
3. The events occur uniformly.
4. The events occur independently.

Principle (3) is equivalent to saying that, over any period chosen at random, there will be a constant mean occurrence of the event, i.e. μ is constant no matter what the sample size.

Principle (4) is equivalent to saying that one occurrence of the event does not influence another occurrence of the event in any way.

The whole process views the occurrence and non-occurrence of the event as the two sole possible kinds of outcome. The examples below represent a few instances in which the Poisson distribution is applicable.

1. The number of cars of a given make which pass a given point on a motorway during a fixed period.
2. The rate of arrival at work of employees in a large factory.
3. The distribution of currants in a very large cake.
4. The distribution of typing errors in a manuscript of 400 pages.

Exercise 3.2

Write down as many situations as you can which fit the Poisson process.

3.5 The mean and variance of the Poisson distribution

In early parts of the chapter it has been assumed that the mean of the distribution (*3.4*) is μ. We will now prove that this is, in fact, so and we will also determine the variance of the distribution.

The sum of all the probabilities of *3.4* is 1. That is:

$$P(0) + P(1) + P(2) + P(3) + \ldots$$
$$= e^{-\mu} + e^{-\mu} \cdot \mu + e^{-\mu} \cdot \frac{1}{2!}\mu^2 + \ldots + e^{-\mu} \cdot \frac{1}{r!}\mu^r + \ldots$$
$$= e^{-\mu}\left[1 + \mu + \frac{1}{2!}\mu^2 + \ldots + \frac{1}{r!}\mu^r + \ldots\right]$$
$$= e^{-\mu} \cdot e^{\mu} = 1$$

* Suppose the mean is k. Then:

$$[P(0) + P(1) + P(2) + \ldots P(r) + \ldots] \times k$$

$$= P(0).0 + P(1).1 + P(2).2 + P(3).3 + \ldots P(r).r + \ldots$$

$$= e^{-\mu}.0 + e^{-\mu}.\mu.1 + e^{-\mu}.\frac{1}{2!}\mu^2.2 + e^{-\mu}.\frac{1}{3!}\mu^3.3 + \ldots e^{-\mu}.\frac{1}{r!}\mu^r.r + \ldots$$

$$= e^{-\mu}.\mu\left[1 + \mu + \frac{1}{2!}\mu^2 + \frac{1}{3!}\mu^3 + \ldots + \frac{1}{(r-1)!}\mu^{r-1} + \ldots\right]$$

$$= e^{-\mu}.\mu.e^{\mu} = e^0.\mu = \mu$$

The Poisson distribution (*3.4*) was determined from the binomial distribution by assuming that the population, n, was large.

The variance of the binomial distribution is npq where $q = 1 - p$. Therefore the variance $= np(1 - p)$. Now $\mu = np$, i.e. $p = \mu/n$. Therefore the variance of the Poisson distribution is:

$$\mu\left(1 - \frac{\mu}{n}\right)$$

$$= \mu - \mu^2/n$$

As $n \to \infty$ the variance tends to μ.

3.6 Bar chart representation of Poisson probabilities

A method of representing Poisson probabilities similar to that used for the binomial distribution may be applied. The following examples represent the bar chart diagrammatic illustration of these probabilities.

Examples

1. Represent in diagrammatic form the probabilities determined in Example 1 in section 3.3.
    Fig. 3.1 represents the first six values of the probabilities in the distribution.
2. Fig. 3.2 represents the distribution in Example 2 in section 3.3. Further values of $P(r)$ are calculated below to obtain a better appreciation of the nature of the distribution.

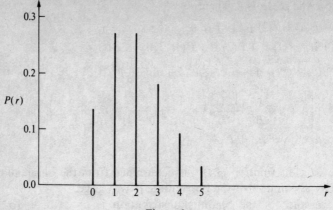

Figure 3.1

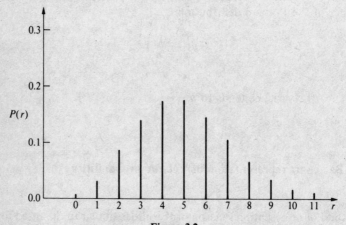

Figure 3.2

$$P(7) = e^{-5} \cdot \frac{1}{7!} \cdot 5^7 = 0.1044449$$

$$P(8) = e^{-5} \cdot \frac{1}{8!} \cdot 5^8 = 0.065278$$

$$P(9) = e^{-5} \cdot \frac{1}{9!} \cdot 5^9 = 0.0362656$$

$$P(10) = e^{-5} \cdot \frac{1}{10!} \cdot 5^{10} = 0.0181328$$

$$P(11) = e^{-5} \cdot \frac{1}{11!} \cdot 5^{11} = 0.0082422$$

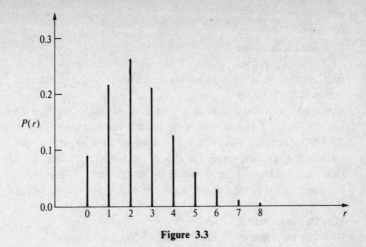

Figure 3.3

3.   Fig. 3.3 is a diagrammatic representation of the distribution in Example 3 in section 3.3. Additional probabilities are calculated below.

$$P(5) = e^{-2.4} \times \frac{1}{5!}(2.4)^5 \quad = 0.0601961$$

$$P(6) = e^{-2.4} \times \frac{1}{6!}(2.4)^6 \quad = 0.0240784$$

$$P(7) = e^{-2.4} \times \frac{1}{7!}(2.4)^7 \quad = 0.0082555$$

$$P(8) = e^{-2.4} \times \frac{1}{8}(2.4)^8 \quad = 0.0024766$$

Note that all the bar charts illustrate distributions which are skewed on the left. This is a characteristic feature of Poisson distributions.

Exercise 3.3

1.   Assuming that the results in Football League matches show that the average number of goals scored per match is 2.2, determine the probabilities of the number of goals scored in any randomly selected match being:
(a)   zero;
(b)   one;

    (c) two;

    (d) more than one;

    (e) less than three;

    (f) more than five.

Represent the distribution in diagrammatic form for goals scored per match from 0 to 10.

*2. Assume that the distribution of personal income of adults within the UK follows a Poisson distribution and that the average income is £4500 per annum. Take a unit of income to be £500 and assume that all incomes are expressed to the nearest £500, so that an annual income of £3500 is represented by seven units. A member of the community is selected at random. Determine the probabilities that the person:

    (a) has no income;

    (b) has an income no more than £5500;

    (c) has an income in excess of £10 000;

    (d) has an income of exactly £6000.

Represent the distribution graphically for incomes from zero to £10 000.

3. On average 1 per cent of a particular manufactured product are defective. In a random sample of 80 items determine the probabilities that:

    (a) none is defective;

    (b) exactly one is defective;

    (c) exactly two are defective;

    (d) more than two are defective.

*4. A new drug is being tested to combat a virulent infection. On average, in addition to curing the infection, it has a serious side effect in 0.02 per cent of cases treated. Assuming that the Poisson distribution is applicable, determine the probabilities that, out of a random sample of 4000:

    (a) exactly four will show side effects;

    (b) not more than six will show side effects;

    (c) more than ten will show side effects.

Represent the distribution diagrammatically for values of side effects from zero to 20 inclusive.

5. On average a machine for filling bottles with shampoo breaks down once every two weeks. Assume the Poisson distribution is applicable. Calculate the probabilities of zero, one, two, three, four and five breakdowns on the machine per week. Determine:

    (a) the probability that the machine breaks down more than twice a week;

(b) the probability that the machine breaks down less than twice a week.

**6. Over a long period a machine turning out metal tags turns out 2 per cent defectives. Calculate the size of a random sample such that the probability that it contains at least two defectives is 0.9 or more.

7. Golf balls are packed 12 to a box. On average, five balls in every 2000 are mis-shapen. Determine the probabilities that a randomly selected box contains:
   (a) exactly one faulty ball;
   (b) at least two faulty balls;
   (c) no faulty balls.

4

Properties of the Normal Curve. Normal distribution probabilities

In Chapter 2 some of the bar chart diagrams produced figures which were symmetrical and bell-shaped. The values of r in those diagrams were discrete and integral. The sum of the ordinates for any of the bar charts, i.e. the sum of the various probabilities, was always 1. The mean was np and the standard deviation was npq. When n is very large such a distribution closely approximates what is called a normal distribution. It is customary to modify the characteristics of such a curve to facilitate calculations. Fig. 4.1 represents the standardized normal curve.

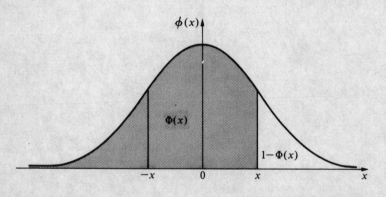

Figure 4.1

4.1 Characteristics of the normal curve

1. $\phi(x) = \dfrac{1}{\sqrt{2\pi}} e^{-\frac{1}{2}x^2}$

2. The mean is $x = 0$.

3. The standard deviation is 1.

4. The variate, x, is continuous and is measured in multiples of the standard deviation.

5. Because the variate is continuous it is no longer possible to use $\phi(x)$ as a measure of $P(x)$, i.e. the probability of x.

6. The area under the curve to the left of the ordinate x is denoted by $\Phi(x)$, and this represents a measure of $P(x)$.

7. The total area under the curve is 1.

8. The area $1 - \Phi(x)$ represents the probability that a given value of the variate will fall outside the value x, i.e. will deviate from the mean by more than $+x$.

9. The area $1 - \Phi(x)$ also represents the probability that a given value of the variate will fall outside the value $-x$.

10. Consequently the probability that a value of the variate will differ from the mean by not more than x (numerically) is represented by the area between the ordinates x and $-x$, i.e. by:

$$1 - 2[1 - \Phi(x)] = 2\Phi(x) - 1$$

Particular values of ϕ and Φ

$$\phi(0) = \frac{1}{\sqrt{2\pi}} = 0.3989423 \qquad \Phi(0) = 0.5000$$

$$\phi(1) = \frac{1}{\sqrt{2\pi}} \cdot e^{-\frac{1}{2}} = 0.2419707 \quad \Phi(1) = 0.8413$$

$$\phi(-1) = \frac{1}{\sqrt{2\pi}} \cdot e^{-1} = 0.2419707 \quad \Phi(-1) = 0.1587$$

$$\phi(2) = \frac{1}{\sqrt{2\pi}} \cdot e^{-2} = 0.053991 \qquad \Phi(2) = 0.9772$$

$$\phi(-2) = \frac{1}{\sqrt{2\pi}} \cdot e^{-2} = 0.053991 \quad \Phi(-2) = 0.0228$$

$$\phi(3) = \frac{1}{\sqrt{2\pi}} e^{-9/2} = 0.0044318 \quad \Phi(3) = 0.9987$$

$$\phi(-3) = \frac{1}{\sqrt{2\pi}} e^{-9/2} = 0.0044318 \quad \Phi(-3) = 0.0013$$

In tables of Φ it is usual to quote values only for the positive range of values of x from 0 to 3 or, perhaps, from 0 to 4. Tables for the normal distribution function are given at the end of the book.

Examples

Using normal distribution tables determine the probabilities for the following normal distributions, given that the mean is equal to zero and the standard deviation is equal to 1.

1. The probability that the variate lies within one standard deviation on each side of the mean.
    The probability is:

    $$2\Phi(1) - 1$$
    $$= 2 \times 0.8413 - 1$$
    $$= 1.6826 - 1$$
    $$= 0.6826$$

Therefore there is a 68.26 per cent probability that a given value of x lies within one standard deviation of the mean zero.

2. The probability that the variate lies within two standard deviations on each side of the mean.
    The probability is:

    $$2\Phi(2) - 1$$
    $$= 2 \times 0.9772 - 1$$
    $$= 1.9544 - 1$$
    $$= 0.9544$$

There is a 95.44 per cent probability that a given value of x lies within two standard deviations of the mean zero.

3. The probability that the variate x lies within three standard deviations on each side of the mean.
    The probability is:

    $$2\Phi(3) - 1$$
    $$= 2 \times 0.9987 - 1$$
    $$= 1.9974 - 1$$
    $$= 0.9974$$

There is a 99.74 per cent probability that a given value of x lies within three standard deviations of the mean zero.

4. Boxes of matches are produced with an average of 50 matches per box and a standard deviation of one match. Determine the probability that a randomly selected box of matches will contain:
    (a) from 49 to 51 matches;
    (b) from 48 to 52 matches;
    (c) fewer than 47 matches;
    (d) more than 52 matches.

    (a) A box with 49 to 51 matches constitutes one with contents lying within one standard deviation of the mean. The probability is:

    $$2\Phi(1) - 1$$
    $$= 0.6826$$

    (b) A box with 48 to 52 matches constitutes one with contents lying within two standard deviations of the mean. The probability is:

    $$2\Phi(2) - 1$$
    $$= 0.9544$$

    (c) A box with fewer than 47 matches constitutes one with contents lying more than three standard deviations below the mean. The probability is:

    $$\Phi(-3)$$
    $$= 1 - \Phi(3)$$
    $$= 1 - 0.9987$$
    $$= 0.0013$$

    (d) A box with more than 52 matches constitutes one with contents lying more than two standard deviations above the mean. The probability is:

    $$1 - \Phi(2)$$
    $$= 1 - 0.9772$$
    $$= 0.0228$$

4.2 Situations appropriate to a normal distribution

Many populations which occur naturally follow normal distribution patterns.

Examples

1. The heights or weights of particular age groups of men, women or children.
2. The lengths of time a particular age group can hold a single breath.
3. The weight of apples per tree in a large orchard of one variety of apple.
4. The volume of detergent in each of a large number of bottles of the same size.
5. The time of reaction of a large number of people to a given signal.
6. The times of arrival on different days of a particular train at a given station.
7. The life of electric lamps of a given make and rating.

Exercise 4.1

Using the normal distribution tables at the end of the book answer the following questions:

1. Determine $\Phi(x)$ when $x = 0.06, 0.52, 1.40, 1.86, 2.09, 2.65, 3.17, 3.58$.
2. Determine the value of $2[1 - \Phi(x)]$ when $x = 0.23, 0.84, 1.07, 1.74, 2.35, 2.94, 2.65, 3.17, 3.58$.
3. Determine the values of $1 - \Phi(x)$ when $x = 1.42, 2.36, 0.57, 3.65$.
4. Determine the values of $\Phi(-x)$ when $x = 0.29, 1.63, 2.74, 3.28$.
5. Determine the values of x for the following values of $\Phi(x)$: $0.5239, 0.5948, 0.6772, 0.7673, 0.9665, 0.99534, 0.4343, 0.3725, 0.2848, 0.1926, 0.0018$.
6. Determine the probability that $x > 1.63$.
7. Determine the probability that $x > 2.39$.
8. Determine the probability that $0 < x < 0.85$.
9. Determine the probability that $0.74 < x < 1.29$.
10. Determine the probability that $1.06 < x < 3.25$.
11. Determine the probability that $x < -1.9$.
12. Determine the probability that $x < -2.32$.
13. Determine the probability that $-2 < x < 0$.
14. Determine the probability that $-1 < x < 1.5$.
15. Determine the probability that $-3.26 < x < -1.04$.
16. Determine the probability that $-2.68 < x < 1.94$.
*17. The probability that x lies between x_1 and x_2 is 0.623. Determine the value of x_2 when $x_1 = -1.32$.
*18. The probability that x is greater than a is 0.4562. Determine the value of a.

*19. The probability that x is less than b is 0.6249. Determine the value of b.

*20. The probability that x is less than c and greater than b is 0.3407. Determine the value of c when the value of b is -1.08.

4.3 The conversion of data from a normal distribution to standardized form

Example 4 in section 4.1 involved data where the mean was not zero, although the standard deviation was one. In such cases, as well as in those where the standard deviation is not one, it is customary to convert the data into standardized form so that the standardized tables may be used directly.

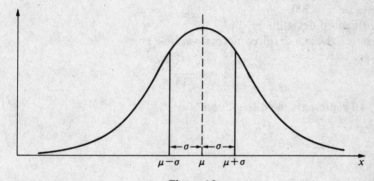

Figure 4.2

Fig. 4.2 represents a normal curve for which the mean is μ and the standard deviation is σ. Then $x - \mu$ represents the deviation of the variable x from the mean. Consequently $|x - \mu|$ represents the numerical value of that deviation. As a result of dividing by σ:

$$\frac{|x - \mu|}{\sigma}$$

represents the transformation of the deviation such that when $|x - \mu|$ takes a value equal to σ the value of the transformed deviation is 1. Therefore the standardized normal variable is:

$$Z = \frac{|x - \mu|}{\sigma}$$

The standardized normal distribution tables may be used with the variable Z. In fact that is the variable which is usually quoted in such tables.

Example

A company producing lard makes up packs with an average weight of 500 g and a standard deviation of 14 g. It is then distributed to retailers in boxes of 250. How many packs in a box may be expected to weigh:
(a)   less than 472 g;
(b)   less than 460 g;
(c)   more than 510 g;
(d)   between 480 g and 530 g;
(e)   at most 520 g?
The mean weight of a pack = 500 g.
Therefore $\mu = 500$.
The standard deviation is = 14.
Therefore the standardized normal variable is:

$$Z = \frac{|x - 500|}{14}$$

Fig. 4.3 represents the distribution.

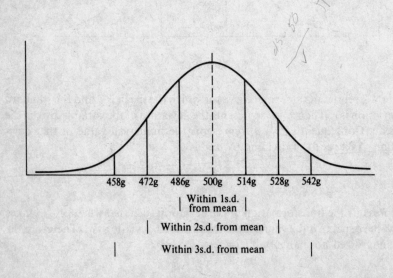

458g   472g   486g   500g   514g   528g   542g

| Within 1s.d.
from mean |

| Within 2s.d. from mean |

| Within 3s.d. from mean |

Figure 4.3

(a) When $x = 472$ the standardized variable takes the value:

$$\frac{472 - 500}{14}$$

$$= -\frac{28}{14}$$

$$= -2$$

i.e. $|Z| = 2$

From the normal tables, for $Z = 2$, $\Phi(Z) = 0.97725$. Then:

$$\Phi(-Z) = 1 - \Phi(Z)$$

$$= 1 - 0.97725$$

$$= 0.02275$$

Therefore $P(x < 472) = 0.2275$. In a box of 250 packs the expected number is:

$$250 \times 0.02275$$

$$= 5.6875$$

$$\approx 6$$

(b) When $x = 460$ the standardized variable takes the value:

$$\frac{460 - 500}{14}$$

$$= -\frac{40}{14}$$

$$= -2.8571429$$

$$\approx -2.86$$

From the tables $\Phi(2.86) = 0.99788$. Therefore:

$$\Phi(-2.86) = 1 - \Phi(2.86)$$

$$= 1 - 0.99788$$

$$= 0.00212$$

The expected number is:

$$250 \times 0.00212$$

$$= 0.53$$

$$\approx 1$$

(c) When $x = 510$ the standardized variable takes the value:

$$\frac{510 - 500}{14}$$

$$= \frac{10}{14}$$

$$= 0.7142857$$

$$\approx 0.71$$

From the tables $\Phi(0.71) = 0.7611$. Therefore:

$$P(x > 510) = [1 - 0.7611]$$

$$= 0.2389$$

The expected number is:

$$250 \times 0.2389$$

$$= 59.725$$

$$\approx 60$$

(d) When $x = 530$ the standardized variable takes the value:

$$\frac{530 - 500}{14}$$

$$= \frac{30}{14}$$

$$= 2.1428571$$

$$\approx 2.14$$

When $x = 480$ the standardized variable takes the value:

$$\frac{480 - 500}{14}$$

$$= -\frac{20}{14}$$

$$= -1.4285714$$

$$\approx -1.43$$

From the tables:

$$\Phi(2.14) = 0.98382$$

$$\Phi(1.43) = 0.9236$$

Therefore:

$$\Phi(-1.43) = 1 - \Phi(1.43)$$
$$= 1 - 0.9236$$
$$= 0.0764$$

Hence:

$$\Phi(2.14) - \Phi(-1.43)$$
$$= 0.98382 - 0.0764$$
$$= 0.90742$$

The expected number is:

$$250 \times 0.90742$$
$$= 226.855$$
$$\approx 227$$

(e) When $x = 520$ the standardized variable takes the value:

$$\frac{520 - 500}{14}$$
$$= \frac{20}{14}$$
$$= 1.4285714$$
$$\approx 1.43$$

From the tables $\Phi(1.43) = 0.9236$. The expected value is:

$$250 \times 0.9236$$
$$= 230.9$$
$$\approx 231$$

Exercise 4.2

1. Bottles of shampoo have a mean capacity of 120 ml and standard deviation of 1.5 ml. In a batch of 500 bottles determine how many bottles are expected to have capacities:
    (a)   more than 123.5 ml;
    (b)   more than 121.5 ml;
    (c)   less than 118 ml;
    (d)   between 116 ml and 124 ml;
    (e)   of at least 119 ml.

2. On average a small mail-order firm receives 256 orders per week by letter. Assume that the pattern of orders follows a normal distribution with a standard deviation of 24. Determine the number of weeks in a period of 50 when the number of expected orders is:
   (a) less than 230;
   (b) more than 285;
   (c) between 240 and 300;
   (d) not less than 220.

3. Sacks of potatoes from a large commercial grower have a mean weight of 25 kg with a standard deviation of 0.5 kg. Determine the percentage of sacks with expected weight:
   (a) more than 26.4 kg;
   (b) less than 24.2 kg;
   (c) between 23.8 kg and 26.6 kg;
   (d) of at least 23.7 kg.

*4. The heights of 9545 Yorkshiremen within the same age group were measured. Their mean height was 1715 mm with a standard deviation of 65 mm. A man from the group was selected at random. Determine the probabilities that his height was:
   (a) over 1800 mm;
   (b) less than 1550 mm;
   (c) between 1550 mm and 1800 mm;
   (d) either more than 1800 mm or less than 1550 mm;
   (e) neither more than 1810 mm nor less than 1540 mm.

*5. A large television rental firm estimates that a particular replacement unit has a mean life of 1250 hours with standard deviation of 125 hours. A new consignment from the factory contains 500 of these units. Determine the number of these units with expected lives of:
   (a) less than 1100 hours;
   (b) between 1150 hours and 1350 hours;
   (c) more than 1400 hours;
   (d) either less than 1050 hours or more than 1450 hours;
   (e) not less than 1075 hours and not more than 1475 hours.

6. A firm producing dry ready-mixed concrete tests a machine set to fill bags with 5 kg. After tests it is discovered that 5.48 per cent of the bags contain more than 5030 g of dry ready-mixed material. Determine:
   (a) the standard deviation of the machine's operation;
   (b) the percentage of bags weighing more than 5050 g;
   (c) the percentage of bags weighing less than 4950 g;
   (d) the percentage of bags weighing between 4950 g and 5050 g.

*7. A manufacturer specifies delivery of a particular model of caravan within 28 days. In practice the mean delivery interval is 23 days with a standard deviation of four days. Determine the percentage of caravans of that model which are:

(a) delivered late;
(b) delivered early;
(c) delivered more than two days early;
(d) delivered more than two days late.

If he wishes to ensure that not more than 5 per cent of the deliveries are late, to what period must he reduce the standard deviation, assuming it is not possible to lower the mean delivery interval?

4.4 The use of normal probability paper

So far it has been assumed that we have been dealing with populations which are normal; that is, their probability patterns follow those of the normal distribution function. However, there are occasions when the characteristics of a particular population are unknown. Frequently we need to know whether the population is normal and we wish to make use of a graphical technique to detect the characteristic. If we plotted the raw data at our disposal and then tried to assess whether or not its shape was the characteristic bell shape of the normal curve, we would find the task difficult.

The one shape we can all recognize easily is that of the straight line. Consequently, where possible, when we are looking for a relationship or pattern which data fit we try to make the decision depend on whether or not the graphical representation of the data is linear or approximates closely to a linear law. In the present context a special kind of graph paper is used to achieve this purpose. It is called normal probability paper. Before we are able to use the paper we may have to adapt the data at our disposal. The following example illustrates the procedure.

Example

The data below represent the frequency of arrival times of employees at a large factory. The times of arrival are given to the nearest minute.

Times (a.m.):	7.50	7.51	7.52	7.53	7.54	
frequency:	6	35	76	148	277	

Times:	7.55	7.56	7.57	7.58	7.59	8.00
Frequency:	408	294	139	69	42	6

Determine whether or not the data fit a normal curve. The raw data give the frequencies with which employees arrive at the works. These data must be modified: the cumulative relative frequencies must be obtained. This may be done in either of two ways:

1. Convert the frequencies to relative frequencies and then calculate the cumulative relative frequencies.
2. Calculate the cumulative frequencies and then convert to cumulative relative frequencies.

From the table above the cumulative frequencies are represented in the table below.

Times:	7.50	7.51	7.52	7.53	7.54	7.55
Cumulative frequencies:	6	41	117	265	542	950

Times:	7.56	7.57	7.58	7.59	8.00
Cumulative frequencies:	1244	1383	1452	1494	1500

One important issue arises in relation to the table above. When, in the first table, the time of arrival is given as 7.58 that represents the mid-interval of the period during which 69 employees arrived. They actually arrived between $7.57\frac{1}{2}$ and $7.58\frac{1}{2}$. Therefore, to be strictly correct, in the cumulative frequency table we should realize that, by $7.58\frac{1}{2}$, 1452 employees had arrived at work. Consequently, the times in that table should all be changed by adding on an extra half minute, and that remains true for the tables which follow. The figures have been left as they are because they look tidier.

In the table below the cumulative relative frequencies have been converted to percentages so that the problem of graphing the results on normal probability paper is made easier.

Times:	7.50	7.51	7.52	7.53	7.54	7.55
Percentage cumulative frequencies:	0.4	2.73	7.8	17.67	36.13	63.33

Times:	7.56	7.57	7.58	7.59	8.00
Percentage cumulative frequencies:	82.93	92.2	96.8	99.6	100.00

The revised data are plotted in Fig. 4.4. Although the points do not lie exactly on a line, there is a close linear relationship between them.

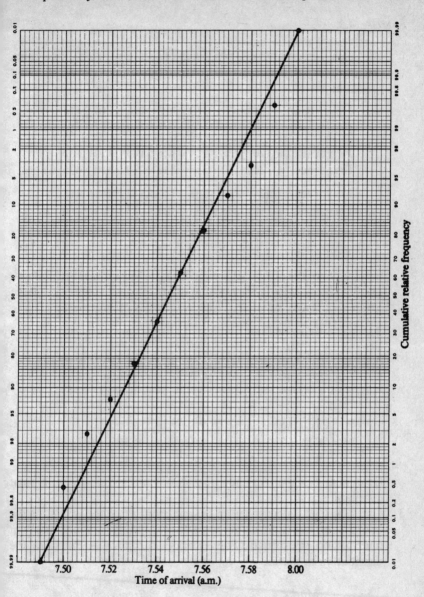

Figure 4.4

Consequently we conclude that these data do conform to an approximate normal distribution.

Exercise 4.3

1. The height, in centimetres, of a group of young boys within the same
    age group are given below, with the frequencies with which these
    heights appeared.

Height (cm):	106	108	110	112	114	116	118
Frequency:	0	1	0	3	9	13	10

Height (cm):	120	122	124	126	128	130	132
Frequency:	20	32	55	130	220	320	425

Height (cm):	134	136	138	140	142	144	146
Frequency:	390	274	144	25	13	1	0

Height (cm):	148	150	152	154	156	158	160
Frequency:	0	3	2	2	0	1	0

    Determine whether or not this approximates to a normal distribution.
2. The data below refer to the rateable values of private houses in a large
    city. The rateable values are given in intervals of roughly £50.

Rateable value (£s)	Frequency (no. of houses)
51–100	2990
101–150	15 020
151–200	29 990
201–250	36 030
251–300	17 080
301–350	11 090
351–400	6010
401–450	1820

    Determine whether or not the data correspond to a normal
    distribution.
3. One thousand students sat an examination. After the results of the
    examination were known they were divided into groups with marks in
    intervals of approximately ten marks. The table below gives the
    distribution of marks.

Marks	Frequency
less than 10	4
10–19	10
20–29	16
30–39	88
40–49	206

50–59	341
60–69	244
70–79	70
80–89	16
90–100	5

Test the data for a normal distribution.

4. A machine produces metal bars with a normal length of 40 cm. One thousand bars are selected and measured correct to the nearest millimetre. The frequencies of their lengths are given in the table below.

Length (cm):	39.5	39.6	39.7	39.8	39.9	40.0
Frequency:	11	14	28	109	182	309

Length (cm):	40.1	40.2	40.3	40.4	40.5
Frequency:	221	74	32	12	8

Test the data for a normal distribution.

5. Test the following data for a normal distribution.

x:	1	2	3	4	5	6	7	8	9	10
y:	1	4	5	12	26	68	121	219	283	375

x:	11	12	13	14	15	16	17	18	19	20
y:	320	319	236	148	50	22	7	3	1	0

5

Point and
Interval Estimates

5.1 Sampling

Early in April 1981 the national census took place. This survey, designed to discover important facts about the population of the UK, takes place only at rare intervals, usually every ten years. It is too costly and lengthy an operation to be undertaken more frequently.

Normally when information about the whole population or a large section of the population is required either by the government or by industry or commerce, a random sample of the population under review is chosen. The information obtained from that sample, if it is typical of the relevant section, is used to help to determine political, financial, economic, industrial and other associated policies of the body or bodies concerned. The methods of selecting the samples do not come within the limits of this chapter. What we are concerned with here is the extent to which measures of samples can be related to the corresponding measures of the population from which the sample is taken. An alternative term for the population is the parent population.

We must be careful to distinguish between the means and standard deviations of the population on the one hand and the means and standard deviations of the sample on the other. To help us to do this it is customary to use different symbols for these two measures in the two cases. The mean of the population is represented by μ and the mean of the sample is designated $\bar{x}$. The standard deviation of the population is represented by σ and that of the sample by s.

When the size of the sample is too small the information obtained from it is not sufficiently reliable. Yet, when the size of the sample is too large, the cost of the survey is unacceptably high. In practice a compromise is made between the two extremes. It is customary to take the value 30 as a dividing line between small and large sample sizes. Where n, the sample size, is greater than 30 we consider the sample to be large and where n, the sample size, is less than 30 we consider the sample small.

In fact a very interesting and important feature emerges concerning the means of random samples of populations.

5.2 The distribution of the means of random samples of a population

The different values of the means, $\bar{x}$, of random samples of a population follow a pattern. They represent an approximately normal distribution. This is true even when the population itself is not normal. For this result to be applicable the sample must be representative of the parent population and the sample size must be large enough.

The proof that the sample means of a population form a normal distribution providing the sample size is large enough is beyond the scope of this book. However, with care and the exercise of much patience and hard work, it is possible to illustrate that the fact is reasonable. When a set of random numbers is not available take ten cards or discs which are numbered from one to ten. Shuffle the cards, select one, note the number and replace the card. Repeat this operation many times and keep a record of all the scores. The data below represent the results of 2000 such operations.

Score:	1	2	3	4	5	6	7	8	9	10
Frequencies:	191	184	194	226	217	210	219	190	233	136

These data produce a mean of 5.473 and a standard deviation of 2.759897. Consequently we shall accept these as the corresponding figures for a population which is very large, i.e. the parent population. This means that:

$$\mu = 5.473$$
$$\sigma = 2.759897$$

Note that a purely theoretical value of μ is 5.5. In addition, the means of samples of 100, 200 and 400 were noted. They are shown below.
Means of sample size 100:

5.12	5.80	5.61	5.98	5.86
5.34	5.18	5.89	5.04	5.19
5.17	5.39	5.27	5.64	5.34
5.58	5.32	5.80	5.83	5.11

Means of sample size 200:

$$5.46 \quad 5.795 \quad 5.6 \quad 5.535 \quad 5.115$$
$$5.28 \quad 5.455 \quad 5.46 \quad 5.56 \quad 5.47$$

Means of sample size 400:

$$5.6275 \quad 5.5675 \quad 5.1975 \quad 5.4575 \quad 5.515$$

Table 5.1 *Standard deviations of distributions*

Sample size	s.d. of means	$\sigma/\sqrt{n}$
100	0.2990502	0.2759897
200	0.1724268	0.1951542
400	0.1487985	0.1379949

In Table 5.1 the standard deviations of these distributions are given and compared with the values of $\sigma/\sqrt{n}$. From the data and calculations above it is clear that the means of samples do tend towards the mean of the parent population as the sample size increases. The calculations for the standard deviation of the means and for the values of $\sigma/\sqrt{n}$ show a reasonably close connection between the two.

5.3 To estimate the mean of a population given the mean of a sample

Providing the sample size is large enough (not merely just greater than 30, but very large), the mean of the sample, $\bar{x}$, is taken as an estimate of the value of μ, the mean of the parent population.

Example

A firm specializes in installing double-glazed window units. From the data of a few hundred installations it determines the mean time for installing a particular model to be 7.5 hours. Here $\bar{x} = 7.5$ hours. Therefore an estimate of μ is also 7.5 hours.

5.4 Confidence limits

In estimating the value of μ to be that of $\bar{x}$, determined from one sample, we need to know the degree to which we can rely on the accuracy of that estimate.

In sampling theory the standard deviation of a sampling distribution, i.e. the standard deviation of the sample means, is called the standard error and it is also designated by the symbol σ_n.

From the early part of the chapter the standard error of $\bar{x}$ is:

$$\sigma_n = \frac{\sigma}{\sqrt{n}}$$

where σ is the standard deviation of the parent population and n is the sample size. For that formula to be applicable σ must be known. When it is not known σ is replaced by s, the standard deviation of the sample. The error introduced is not serious when n is large.

5.5 The central limit theorem

Because the sample means represent an approximate normal distribution an important theorem follows. This is the central limit theorem. It states that if a random sample of size n is taken from a population whose mean and variance are μ and σ^2 respectively, then:

$$\frac{\bar{x} - \mu}{\sigma/\sqrt{n}}$$

tends to the normal distribution as n increases.

Examples

1. A large population has a mean of 32.7 and standard deviation of 4.23. Samples of different sizes are taken. Determine the standard deviations of the means of samples of the following sizes: (a) 40; (b) 50; (c) 100; (d) 200.
   The calculations are given below.

   (a) $\sigma_n = \dfrac{4.23}{\sqrt{40}} = 0.6688217 \approx 0.67$

(b) $\sigma_n = \dfrac{4.23}{\sqrt{50}} = 0.5982123 \approx 0.60$

(c) $\sigma_n = \dfrac{4.23}{\sqrt{100}} = 0.423 \qquad \approx 0.42$

(d) $\sigma_n = \dfrac{4.23}{\sqrt{200}} = 0.2991062 \approx 0.30$

Fig. 5.1 represents a sketch of the distribution of the means of sample sizes 40, 50, 100 and 200. Note how the standard deviation of the means of samples decreases as the sample size increases. That indicates that the larger the sample the more closely the sample means will cluster round the mean of the population. Therefore the larger the sample size the more confident we can be that the sample mean will closely approximate the population mean.

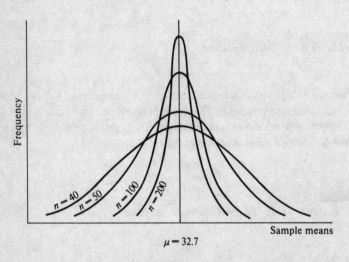

Figure 5.1

2. The mean tensile strength of a large number of metal blocks is 403.6 N/mm², with a standard deviation of 17.8 N/mm². Random samples of 50 blocks are taken. Determine:
   (a) the standard deviation of the sample means;
   (b) the probability that the mean tensile strength of a sample lies between 405 N/mm² and 402 N/mm²;

(c) the probability that the mean tensile strength of 4 per cent of the samples is less than 401 N/mm^2.

From the data $\mu = 403.6$ and $\sigma = 17.8$.

(a) The standard deviation of the sample means is:

$$\sigma_n = \frac{\sigma}{\sqrt{n}}$$

$$= \frac{17.8}{\sqrt{50}}$$

$$= 2.5173001$$

$$\approx 2.517$$

(b) The distribution $\dfrac{\bar{x} - 403.6}{2.5173001}$, where $\bar{x}$ is the mean of the sample, is approximately normal. When $\bar{x} = 405$:

$$\frac{\bar{x} - \mu}{\sigma/\sqrt{n}} = \frac{405 - 403.6}{2.5173001}$$

$$= 0.5561514$$

$$\approx 0.56$$

When $\bar{x} = 402$:

$$\frac{\bar{x} - \mu}{\sigma/\sqrt{n}} = \frac{402 - 403.6}{2.5173001}$$

$$= -0.6356016$$

$$\approx -0.64$$

Then:

$$\Phi(0.56) - \Phi(-0.64)$$

$$= 0.7123 - [1 - \Phi(0.64)]$$

$$= 0.7123 - [1 - 0.7389]$$

$$= 0.7123 + 0.7389 - 1$$

$$= 0.4512$$

The probability that the mean tensile strength of a sample lies between the given limits is 0.4512. Therefore approximately 45 per cent of the samples have a mean tensile strength lying between the given limits.

(c)   When $\bar{x} = 401$:

$$\frac{\bar{x} - \mu}{\sigma/\sqrt{n}} = \frac{401 - 403.6}{2.5173001}$$

$$= -\frac{2.6}{2.5173001}$$

$$= -1.0328526$$

$$\approx -1.03$$

$$\Phi(-1.03) = 1 - \Phi(1.03)$$

$$= 1 - 0.8485$$

$$= 0.1515$$

Therefore 15 per cent of the samples will have a mean tensile strength less than 401 N/mm^2. It is, therefore, highly probable that 4 per cent of the samples will have a mean tensile strength less than 401 N/mm^2.

Exercise 5.1

For the data given below determine the standard deviations of the means of the sample sizes indicated.

1. $\sigma = 10.42$; $n = 30, 40, 50, 100, 150$
2. $\sigma = 19.78$; $n = 50, 100, 200, 500$
3. $\mu = 206.5$; $\sigma = 23.71$; $n = 50, 60, 70, 80, 90, 100$

Using the central limit theorem write down the standardized normal distributions of the means of samples of the following populations with the characteristics given below.

4. $\mu = 98.7$; $\sigma = 0.42$; $n = 30, 40, 50, 100, 150$
5. $\mu = 202.6$; $\sigma = 19.78$; $n = 50, 100, 200, 500$
6. $\mu = 206.5$; $\sigma = 23.71$; $n = 50, 60, 70, 80, 90, 100$
*7. General fertilizer is a mixture of a number of basic fertilizers. In quantity it has a mean content of nitrogen which is 17.3 per cent of the bulk with a standard deviation of 0.92 per cent. It is packed into plastic bags each containing 10 kg. The plastic bags are dispatched in batches of 50. Determine:
  (a)   the standard deviation of the mean nitrogen content of the batches;
  (b)   the probability that the mean nitrogen content of a batch lies between 17.1 per cent and 17.4 per cent;

(c) the probability that the mean nitrogen content of a batch is less than 17.0 per cent.

*8. Ten thousand manual workers in a particular industry have a mean monthly wage of £270 with a standard deviation of £38. Random samples of 500 workers are drawn from the population. Determine:
   (a) the standard error of the means of the samples;
   (b) the probability that the mean of a particular sample is less than £230;
   (c) the probability that the mean of a particular sample lies between £260 and £280.

*9. A machine produces plastic pegs to a specification of 50 mm with a standard deviation of 0.5 mm. A random sample of 50 pegs is taken. Determine:
   (a) the standard error of the means of such samples;
   (b) the probability that the mean of such a sample is 50.08 mm or more;
   (c) the probability that the mean of the sample does not lie between 49.94 mm and 50.06 mm.

5.6 Confidence limits

Fig. 5.2 represents the standardized normal distribution. The unshaded area in Fig. 5.2 is symmetrically divided by the line of the standardized mean, 0. It represents 95 per cent of the total area under the curve. Therefore

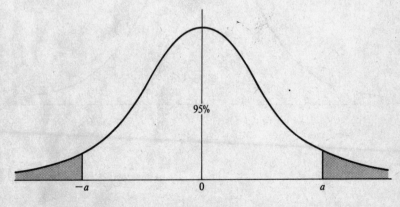

Figure 5.2

each shaded area represents 2.5 per cent of the total area. Consequently the whole of the area to the left of *a* is 97.5 per cent of the total area. Therefore:

$$\Phi(a) = 0.975$$

From the normal distribution tables:

$$\Phi(1.96) = 0.975$$

Therefore $a = 1.96$ and $-a = -1.96$. The area included between upper and lower limits of 1.96 standard deviations from the mean is 95 per cent of the whole. Consequently the probability that the value of a sample mean lies between 1.96 s.d. from the mean is 0.95. When this fact is expressed in terms of confidence limits we obtain the following. The 95 per cent confidence limits for a sample mean $\bar{x}$ are given by:

$$\mu \pm 1.96 \text{ s.d.}$$

When the sample mean is known corresponding confidence limits for a population mean are:

$$\bar{x} \pm 1.96 \text{ s.d.}$$

Fig. 5.3 represents a corresponding situation when the two limits are *b* and $-b$ and when they enclose an area which is 99 per cent of the total area under the curve. Each shaded area represents 0.5 per cent of the total area.

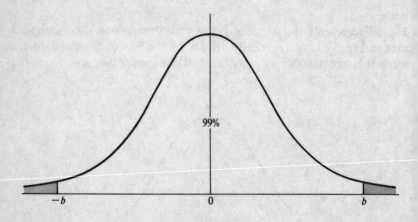

Figure 5.3

Therefore:

$$\Phi(b) = 0.99 + 0.005$$
$$= 0.995$$

From the normal distribution tables:

$$\Phi(2.576) = 0.995$$

Therefore $b = 2.576$ and $-b = -2.576$. In relation to confidence limits these values indicate that 99 per cent confidence limits for the sample mean are given by:

$$\mu \pm 2.576 \text{ s.d.}$$

Corresponding confidence limits for a population mean are:

$$\bar{x} \pm 2.576 \text{ s.d.}$$

Similarly, Fig. 5.4 may be used to help to determine 99.9 per cent confidence limits. From Fig. 5.4:

$$\Phi(c) = 0.9995$$

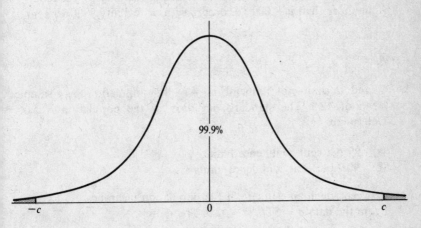

99.9%

$-c$ 0 c

Figure 5.4

From the tables:

$$\Phi(3.3) \approx 0.9995$$

Then $c = 3.3$ and $-c = -3.3$. The 99.9 per cent confidence limits for the sample mean are given by:

$$\mu \pm 3.3 \text{ s.d.}$$

Those for a population mean are:

$$\bar{x} \pm 3.3 \text{ s.d.}$$

In similar fashion 99.99 per cent confidence limits for the sample mean are given by:

$$\mu \pm 3.9 \text{ s.d.}$$

Those for a population mean are:

$$\bar{x} \pm 3.9 \text{ s.d.}$$

So far the confidence limits established refer to the confidence that a sample mean lies between upper and lower limits which are equal amounts above and below a central value. Suppose we wish to consider 95 per cent of the sample means when the excluded 5 per cent lie at one end only, either at the top or the bottom of the range. The normal distribution tables indicate that:

a value less than $\mu + 1.645$ s.d. occurs with probability 95 per cent
a value less than $\mu + 2.33$ s.d. occurs with probability 99 per cent
a value more than $\mu + 3.09$ s.d. occurs with probability 99.9 per cent

Examples

1. A random sample of 70 members of a large population has a sample mean of 42.9. The standard deviation of the population is 5.06. Determine:

   (a) 99 per cent confidence limits;
   (b) 99.9 per cent confidence limits

   between which an estimate of the population mean lies.
   From the data $\sigma = 5.06$, $\bar{x} = 42.9$. Therefore:

   $$\sigma_n = \frac{\sigma}{\sqrt{n}} = \frac{5.06}{\sqrt{70}}$$
   $$= 0.6047857$$
   $$\approx 0.605$$

   Estimated $\mu = 42.9$
   (a) Therefore, with 99 per cent confidence, we can say the mean of the population lies between:

   $$42.9 \pm 2.576 \times 0.605$$
   $$\text{i.e. between } 42.9 \pm 1.55848$$
   $$\text{or between } 41.34152 \text{ and } 44.45848$$
   $$\text{i.e. between } 41.34 \text{ and } 44.46$$

(b) With 99.9 per cent confidence we can say the mean of the population lies between:

$$42.9 \pm 3.29 \times 0.605$$

i.e. between 42.9 ± 1.99045

or between 40.90955 and 44.80955

i.e. between 40.91 and 44.81

2. A machine fills bottles with aftershave lotion. A random sample of 100 bottles has a sample mean of 145 ml content with a standard deviation of 7.3 ml. Determine:

(a) an upper 95 per cent confidence limit;
(b) a lower 99 per cent confidence limit
for the estimated value of the population mean.
From the data $\bar{x} = 145$; estimated $\mu = 145$; $s = 7.3$ and σ is unknown. Therefore an estimated value of σ is 7.3.

(a) With 95 per cent confidence we can say that the population mean is less than:

$$145 + 1.645 \times \frac{7.3}{10}$$

$$= 145 + 1.20085$$

$$= 146.20085 \approx 146.2$$

We can be 95 per cent certain that the population mean is less than 146.2 ml.

(b) With 99 per cent confidence we can say that the population mean is greater than:

$$145 - 2.33 \times \frac{7.3}{10}$$

$$= 145 - 1.7009$$

$$= 143.2991 \approx 143.3$$

We can be 99 per cent certain that the population mean is more than 143.3 ml.

Exercise 5.2

When random samples taken from a large population have the characteristics given below, determine the required confidence limits of the population means.

1. $n = 50$; $\bar{x} = 33.6$; $\sigma = 4.82$. Between upper and lower limits:
   - (a) with 95 per cent confidence;
   - (b) with 99 per cent confidence;
   - (c) with 99.9 per cent confidence.
2. $n = 100$; $\bar{x} = 132.7$; $\sigma = 21.4$. Between upper and lower limits with:
   - (a) 95 per cent confidence;
   - (b) 99 per cent confidence.
3. $n = 80$; $\bar{x} = 302.8$; $\sigma = 29.6$. Below a limit with
   - (a) 95 per cent confidence;
   - (b) 99.9 per cent confidence.
4. $n = 60$; $\bar{x} = 86.05$; $\sigma = 7.93$. Above a limit with
   - (a) 95 per cent confidence;
   - (b) 99.9 per cent confidence.
5. $n = 200$; $\bar{x} = 39.6$; $s = 4.14$. Between upper and lower limits with:
   - (a) 99 per cent confidence;
   - (b) 99.99 per cent confidence.
*6. A random sample of 200 men aged between 20 and 30 is selected from a population in a very large city. Their mean height is 174 cm with a standard deviation of 5.4 cm. Determine:
   - (a) the upper and lower limits for the population mean with 95 per cent confidence;
   - (b) the lower limit for the population mean with 99 per cent confidence;
   - (c) the upper limit for the population mean with 99.9 per cent confidence.

   In this instance the relevant population is the whole group of men in the city aged between 20 and 30.
*7. A random sample of 600 year-old starlings have a wing span of 29.5 cm with a standard deviation of 3.68 cm.
   - (a) With 95 per cent confidence limits estimate the mean wing span of year-old starlings.
   - (b) Determine the size of sample producing the same sample mean and standard deviation if the 95 per cent confidence interval of the year-old starling population was (29.5 ± 0.13) cm.

*5.7 Unbiased estimate of the population mean and variance from sample data

When the mean of a random sample of a large population has been determined, that value is also used as an estimate of the population mean. The process was described earlier in the chapter. The reliability of the estimate depends on the size of the sample. The larger the sample size the more reliable the sample mean is as an estimate of the population mean.

Also, in earlier parts of the chapter, an estimate of the variance of the population was taken to be the variance of the sample. The argument which follows demonstrates why that estimate is not accurate enough. It also paves the way for a concept which is an integral part of other important distributions used for determining confidence intervals.

Suppose the size of a sample of a large population is n. There are n values $x_1, x_2, x_3, \ldots x_n$ which measure the relevant characteristics of the n members of the sample. They are independent of each other. The mean of the population, $\bar{x}$, is:

$$\frac{1}{n}(x_1 + x_2 + x_3 + \ldots + x_r + \ldots x_n)$$

$$= \frac{1}{n} \sum_{r=1}^{n} x_r$$

Therefore $\bar{x}$ is not independent of the x_rs, and this value $\bar{x}$ is taken to be the *mean of the population* no matter what the sample size is or what particular sample is chosen. In other words, we are treating it as a constant. Therefore we treat the sum of the nx_rs as a constant. Consequently, in this respect, they cannot all be independent. Therefore $(n-1)$ of them may be regarded as independent and the remaining one is dependent on them. It follows that the system has $(n-1)$ degrees of freedom.

The variance of the sample is calculated from the values of $(x_1 - \bar{x})$, $(x_2 - \bar{x})$, $(x_3 - \bar{x}) \ldots (x_n - \bar{x})$ or the values of $(x_r - \bar{x})$ for values of r from 1 to n. Suppose s^2 is the variance of the sample and σ^2 is the variance of the population. Then the best unbiased estimate of the population variance is:

$$\frac{s^2 n}{n-1}$$

Sometimes this is written $\hat{\sigma}^2$ to distinguish it from σ^2, the true variance of the population.

Examples

1. Suppose the following data apply to a random sample of a large population: $n = 80$; $x = 326$; $s^2 = 36$.
   Since n is large we take $\sigma^2 = s^2 = 36$. Therefore:

$$\hat{\sigma}^2 = \frac{36 \times 80}{79}$$

$$= 36.455696$$

Therefore $\sigma = 6$ and $\hat{\sigma} = 6.0378553$. The value 36 is regarded as the best unbiased estimate of σ^2.

2. Suppose the following data apply to a random sample of a large population: $n = 15$; $\bar{x} = 326$; $s^2 = 36$.
   By the revised formula for the best unbiased estimate of σ^2:

$$\hat{\sigma}^2 = \frac{36 \times 15}{14}$$

$$= 38.571429$$

$$\hat{\sigma} = 6.21059$$

From the two examples above we can detect the influence of the factor $n/(n-1)$ on the calculation of the variance of the population. When n is large the factor does not have a significant effect; when n is small the factor does have an appreciable effect. A sample is taken to be large when the factor $n/(n-1)$ approximates to the value 1. When $n = 30$, $n/(n-1) = 30/29 = 1.0344828$. There is general agreement that a sample is large when $n > 30$.

Exercise 5.3

With the data listed below determine the best unbiased estimates of the mean and variance of the populations from which the samples are taken.

1. $n = 100$; $\bar{x} = 375$; $s^2 = 49$
2. $n = 200$; $\bar{x} = 582$; $s^2 = 64$
3. $n = 500$; $\bar{x} = 26.9$; $s^2 = 4.41$
4. $n = 50$; $\bar{x} = 106.8$; $s = 3.5$
5. $n = 35$; $\bar{x} = 1325$; $s^2 = 805$
6. $n = 20$; $\bar{x} = 1325$; $s^2 = 805$
7. $n = 15$; $\bar{x} = 106.8$; $s = 3.5$
8. $n = 10$; $\bar{x} = 26.9$; $s^2 = 4.41$
9. $n = 8$; $\bar{x} = 582$; $s^2 = 64$
10. $n = 12$; $\bar{x} = 375$; $s^2 = 49$

5.8 The *t*-distribution

When sample sizes are small the sample means are not normally distributed. Therefore confidence limits for estimating the population mean from a sample mean must not be calculated using the tables of the normal distribution function. An alternative table is called the t-distribution, due to W. S. Gosset, who wrote under the pen-name of 'Student'. The distribution is often called the Student t distribution for that reason. The formula for determining t is:

$$t = \frac{(\bar{x} - \mu)}{s}\sqrt{v+1}$$

where v is the number of degrees of freedom and $v = n - 1$. Therefore:

$$\bar{x} - \mu = t\frac{s}{\sqrt{n}}$$

$$\text{giving} \quad \bar{x} = \mu + t\frac{s}{\sqrt{n}}$$

There are different t-curves according to the number of degrees of freedom. They are similar in shape to the normal curves, but flatter. Fig. 5.5 represents a typical t-curve. The t-distribution table gives values of $Pr(|t| > T)$ represented by the shaded area under the curve. To determine a 95 per cent confidence interval T must take a numerical value which will be exceeded by chance only 5 per cent of the time. This means that the sum of the areas to the right of T and to the left of $-T$ represents 5 per cent of the total area under the curve. In this event those values of T are often written $\pm t_{0.025}$. Therefore the 95 per cent confidence limits of the population mean are:

$$\bar{x} + t_{0.025} \times \frac{s}{\sqrt{n}} \quad \text{and} \quad \bar{x} - t_{0.025} \times \frac{s}{\sqrt{n}}$$

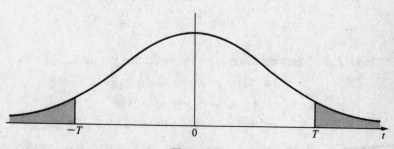

Figure 5.5

The 99 per cent confidence limits are:

$$\bar{x} + t_{0.005} \times \frac{s}{\sqrt{n}} \quad \text{and} \quad \bar{x} - t_{0.005} \times \frac{s}{\sqrt{n}}$$

The 99.8 per cent confidence limits are:

$$\bar{x} + t_{0.001} \times \frac{s}{\sqrt{n}} \text{ and } \bar{x} - t_{0.001} \times \frac{s}{\sqrt{n}}$$

Examples

1. Determine 95 per cent confidence interval limits for the estimated mean of a large population from a random sample of 15 with a mean of 23.5 and a variance of 6.24.

    The number of degrees of freedom = 14. Therefore the t-tables give $t_{0.025} = 2.145$. The 95 per cent confidence limits for the population mean are:

    $$23.5 \pm 2.145 \times \sqrt{\frac{6.24}{15}}$$

    $$= 23.5 \pm 1.3834834$$

    $$= 24.883483 \text{ and } 22.116517$$

    or approximately 24.88 and 22.12

2. A sample of 12 plastic plugs is taken from a large consignment. Their lengths, in centimetres, are: 3.88, 4.02, 4.22, 4.26, 4.42, 4.18, 3.99, 4.06, 4.16, 4.02, 3.90, 3.78. Determine:
    (a) 95 per cent confidence interval limits for the population mean;
    (b) 99 per cent confidence interval limits for the population mean.
    From the data: $\bar{x} = 4.0741667$; $v = 11$; $s = 0.1813314$;

    $$t_{0.025} = 2.201; \ t_{0.005} = 3.106; \ \frac{s}{\sqrt{n}} = \frac{0.1813314}{\sqrt{12}} = 0.0523459$$

    (a) The 95 per cent confidence limits for the population mean are:

    $$4.0741667 \pm 2.201 \times 0.0523459$$

    $$\text{i.e. } 4.18938 \text{ and } 3.9589534$$

    or approximately 4.190 and 3.959

    (b) The 99 per cent confidence limits for the population mean are:

    $$4.0741667 \pm 3.106 \times 0.0523459$$

    $$\text{i.e. } 4.2367531 \text{ and } 3.9115803$$

    or approximately 4.237 and 3.912

Exercise 5.4

For the random samples with data listed below calculate:
(a)   95 per cent confidence interval limits for the population mean;
(b)   99 per cent confidence interval limits for the population mean.

1. $n = 10$; $\bar{x} = 37.5$; $s = 4.45$
2. $n = 15$; $\bar{x} = 106.8$; $s^2 = 90.5$
3. $n = 20$; $\bar{x} = 4218$; $s = 20.3$
4. x takes the values 14.5, 13.2, 15.1, 13.8, 14.1
5. x takes the values 3.26, 3.91, 3.65, 3.43, 3.51, 3.35
6. x takes the values 32.7, 30.3, 31.4, 33.0, 32.0, 31.3, 34.0, 31.2
7. x takes the values 3.41, 3.72, 4.13, 3.44, 4.28, 3.89, 4.18, 4.00, 3.73, 3.93
*8.   Ten random shots from a rifle are observed and their muzzle velocities determined. Measured in metres per second, they are: 3.95, 4.00, 4.03, 3.98, 4.08, 3.96, 4.01, 4.06, 3.99, 4.05. Calculate, for the estimated mean muzzle velocity:
(a)   95 per cent confidence interval limits;
(b)   99 per cent confidence interval limits;
(c)   99.8 per cent confidence interval limits.
*9.   Equal amounts of fibreglass compound are taken from six randomly selected similar packets. Under similar conditions the setting times in minutes for these samples are recorded. They are 35, 43, 39, 36, 47 and 41 minutes respectively. Determine, for the estimated mean setting times of this compound:
(a)   95 per cent confidence interval limits;
(b)   99 per cent confidence interval limits;
(c)   99.8 per cent confidence interval limits.
*10.   From a total of 1000 students sitting a mathematics examination who are listed in alphabetical order, the mark of every 100th student is noted. The marks are 83, 74, 41, 60, 52, 61, 48, 57, 47, 62. Determine, of the mean mark in mathematics of all 1000 students:
(a)   the 95 per cent confidence interval limits;
(b)   the 99 per cent confidence interval limits
*11.   On 20 randomly selected days in a year the minimum temperatures in an office are observed. The temperatures, in degrees Fahrenheit, are 61, 61, 63, 59, 64, 68, 70, 65, 69, 71, 64, 71, 68, 62, 63, 60, 72, 66, 64, 67. Determine, for the mean minimum temperature throughout the year:
(a)   the 99 per cent confidence interval limits;
(b)   the 99.8 per cent confidence interval limits.

6
Simple Hypothesis Tests

6.1 The use of hypotheses

A hypothesis is variously defined as:

1. A supposition or an assumption.
2. A proposition assumed to be true for the sake of argument.
3. A theory which may be proved or disproved by reference to facts.
4. A provisional explanation of some occurrence.

In statistics the term 'hypothesis' is used widely, especially in circumstances where, in relation to a given population, e.g. a production process, a weather pattern, or a sales record, a certain situation or standard has existed. Because of changes in circumstances it is necessary to detect any alteration in those standards.

Null hypothesis

Initially the statistician often begins with the assumption (hypothesis) that there has been no alteration in standards. Such a hypothesis, because it assumes no change, is called a null hypothesis. This null hypothesis is either true or false. The purposes of the various tests are to determine to what degree we feel confident in rejecting the null hypothesis or not rejecting it. Note that although our evidence might lead us to reject a null hypothesis it never leads us, or seldom leads us, to accept it. All we can say on the latter score is that the evidence does not convince us that we should accept the null hypothesis, so judgement is reserved.

Example

Suppose the monthly sunshine figures for June at a particular seaside resort were collected over a long period and the mean figure was 210 hours. Subsequently the June figures for six years, chosen at random, were 204, 219, 198, 223, 163 and 235 hours. Determine whether there is any evidence to suggest that the mean monthly sunshine figure for June at that resort is changing.

Take the null hypothesis, represented symbolically by H_0, to be that there is no change in the mean figure. This is abbreviated to:

$$H_0: \mu = 210$$

At the same time as we postulate the null hypothesis we also propound an alternative hypothesis, H_1, so that, should the null hypothesis be rejected, the alternative hypothesis is accepted. There are three procedures for expressing the alternative hypothesis:

1. That there is an increase in the mean figure; symbolically $H_1: \mu > 210$.
2. That there is a decrease in the mean figure; symbolically $H_1: \mu < 210$.
3. That there is a change in the mean figure; symbolically $H_1: \mu \neq 210$.

From the data above $\bar{x} = 207$; $s = 25.337719$. Take:

$$H_0: \mu = 210$$
$$H_1: \mu < 210$$

Since the sample size is small the t-distribution is used.

$$t = \frac{(\bar{x} - \mu)\sqrt{n}}{s}$$
$$= \frac{(207 - 210)\sqrt{6}}{25.337719}$$
$$= -0.2900209 \approx -0.290$$

Because H_1 is taken to be that $\mu < 210$, a one-tailed test is applied to the t-tables. From the tables, for five degrees of freedom:

$$t_{0.25} = 0.727$$

Fig. 6.1 represents the relationship between the t-value obtained for the sample mean 207 and certain significant t-values of the t-tables for five degrees of freedom. For a calculated negative t-value to be significant at any percentage level it must lie to the left of that particular percentage t-value, i.e. be numerically larger. In this case the calculated t-value is not even significant at the 25 per cent significance level, and this means that the

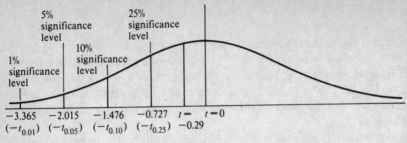

Figure 6.1

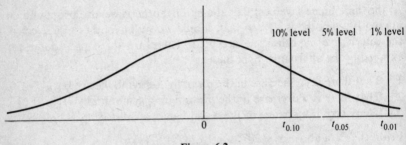

Figure 6.2

probability of such a value occurring by chance is greater than 25 per cent. That is not a rare occurrence, so we do not reject the null hypothesis. The figure for a positive t-value corresponding to Fig. 6.1 is Fig. 6.2. When t lies between $t_{0.10}$ and $t_{0.05}$ the result is significant at the 10 per cent significance level but not at the 5 per cent significance level. In other words, the probability of the result occurring by chance is a value between 5 per cent and 10 per cent: not unlikely. When t lies between $t_{0.05}$ and $t_{0.01}$ the result is significant at the 5 per cent significance level but not at the 1 per cent significance level. In other words, the probability of the result occurring by chance is a value between 1 per cent and 5 per cent: i.e. less likely. When t lies to the right of $t_{0.01}$, i.e. greater, the result is significant at the 1 per cent level. The probability of the result occurring by chance is less than 1 per cent: i.e. unlikely. Figs 6.1 and 6.2 are applicable to one-tailed tests. Where the alternative hypothesis under test is (3) in the June sunshine figures example, a two-tailed test must be applied, and Fig. 6.3 is appropriate. When t lies between either $t_{0.025}$ and $t_{0.005}$ or $-t_{0.025}$ and $-t_{0.005}$ the result is significant at the 5 per cent level but not at the 1 per cent level: i.e. fairly significant. When t lies to the right of $t_{0.005}$ or to the left of $-t_{0.005}$ the result is significant at the 1 per cent level: i.e. significant. When the

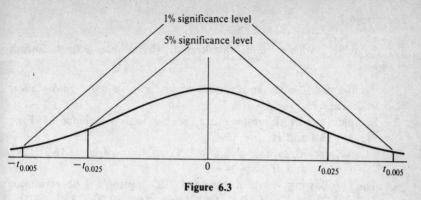

Figure 6.3

probability of an event occurring is 5 per cent then that is not necessarily so significant that the null hypothesis is rejected. It is evidence that we must be on our guard for possible change taking place. When the probability of the event occurring is 1 per cent then that is significant and should normally indicate that we should be ready to reject the null hypothesis.

In the above example, had the mean of the six observations been 174 hours with the same standard deviation, then:

$$t = \frac{|174 - 210|\sqrt{6}}{25.337719}$$

$$= 3.4802513 \approx 3.480$$

$$t_{0.01} = 3.365$$

Consequently the occurrence of a sample mean monthly figure of 174 hours is scarcely likely to appear by chance. On that evidence we would reject the null hypothesis and conclude that this sample indicated a decrease in the mean monthly figure for June. In fact, for this problem, the decision rules are:

1. Accept H_0 and reject H_1 if $t < 1.771$.
2. Do not reject H_0 and do not accept H_1 if $1.771 < t < 2.650$.
3. Reject H_0 and accept H_1 if $t > 2.650$.

Here H_1 is not accepted and that decision is based on the 1 per cent significance level. Decision rules depend very much on the individual who is making the decision, as well as on the circumstances in which those decisions are made. By some, rule (2) above might be changed to read:

2. Reject H_0 and accept H_1 if $1.771 < t < 2.650$.

In that event acceptance would be based on the 5 per cent significance level.

Exercise 6.1

Test the null hypotheses against the alternative hypotheses in the following examples.

1. Sample size 10 with sample mean 4.8 and sample standard deviation 0.52. Take $H_0:\mu = 4.6$ and $H_1:\mu \neq 4.6$.
2. Sample size 18 with sample mean 15.9 and sample variance 4.1. Take $H_0:\mu = 16.4$ and $H_1:\mu < 16.4$.
3. Sample size 20, sample mean 36.5. Population variance 16.2. Take $H_0:\mu = 34.9$ and $H_1:\mu > 34.9$.
4. The following data represent 20 sample observations:
   6.50; 6.50; 5.32; 5.45; 5.08; 5.16; 4.99; 7.29; 4.47; 4.43;
   5.39; 5.16; 4.93; 5.11; 5.75; 5.32; 5.16; 4.47; 5.16; 5.75.
   Test the hypothesis that the mean of the population is 5.1. Give 99 per cent confidence limits for the mean.
5. The following values come from a sample of a large population:
   24.9; 26.8; 31.7; 23.5; 32.6; 28.7; 25.2; 30.1. Test the hypothesis that the population mean is 30 against the alternative hypothesis that it is not 30.
*6. The links in metal chains of a standard design have mean breaking strength of 18 000 N with a standard deviation of 1000 N. It is believed that the addition of an extra layer of tungsten compound to the surface of the links will increase the mean breaking strength. A test of a sample of 25 links shows a mean breaking strength of 20 000 N. At the 1 per cent significance level does this bear out the conclusion that the mean breaking strength has been increased by the added process?
*7. A machine which produces nails is set to manufacture them with a mean length of 3.000 cm. A random sample of 12 nails are found to possess the following lengths: 3.002; 3.008; 2.996; 3.006; 3.008; 3.006; 2.998; 3.010; 3.008; 3.006; 3.008; 3.010 cm. Does the evidence of the sample warrant the conclusion that the mean length is 3.000 cm?

6.2 The critical region

Example

A random sample of size ten is taken from a large population with an estimated mean of 23.7 and an estimated standard deviation of 2.94. Use 95 per cent and 99 per cent confidence limits to determine the critical regions in order that the sample mean indicates an increase in the population mean. For sample size ten, using the t-distribution table for a one-tailed test:

at the 5 per cent significance level, $t = 1.833$
at the 1 per cent significance level, $t = 2.821$

Therefore, when $\bar{x}_1$ represents the sample mean, at the 5 per cent significance level:

$$1.833 = \frac{(\bar{x}_1 - 23.7)}{2.7891289} \sqrt{10} \text{ since}$$

$$s = 2.94 \bigg/ \sqrt{\frac{n}{n-1}} = 2.94 \bigg/ \sqrt{\frac{10}{9}} = 2.7891289$$

$$\text{Therefore } \bar{x}_1 = 23.7 + \frac{1.833 \times 2.7891289}{\sqrt{10}}$$

$$= 23.7 + 1.616706$$

$$= 25.316706$$

$$\approx 25.3$$

At the 1 per cent significance level $\bar{x}_2$, the sample mean, is given by:

$$2.821 = \frac{(\bar{x}_2 - 23.7)}{2.7891289} \sqrt{10}$$

Therefore
$$\bar{x}_2 = 23.7 + \frac{2.821 \times 2.7891289}{\sqrt{10}}$$

$$= 23.7 + 2.488122$$

$$= 26.188122$$

$$\approx 26.2$$

Consequently, when the sample mean reaches the value 25.3 this is taken to be a warning that an increase in the population mean might have taken place. Should the sample mean reach the value 26.2 then we can be 99 per cent certain that, in fact, an increase has taken place. Consequently the attitude towards an increase in the sample mean would be explained by the following decision rules:

1. When the sample mean takes a value greater than 25.3 and less than 26.2 that indicates a possible increase in the population mean.
2. When the sample mean takes a value greater than 26.2 that almost certainly indicates an increase in the population mean.

Definitions

Suppose μ is the estimated mean of a large population, that n is the size of a small random sample from the population, that $t_{0.05}$ is the t-value for $(n-1)$ degrees of freedom at the 5 per cent one-tailed significance level, that $t_{0.01}$ is the t-value for $(n-1)$ degrees of freedom at the 1 per cent one-tailed significance level and that s is the standard deviation of the sample. To determine $\bar{x}_1$, the critical value of $\bar{x}$ at the 5 per cent significance level, and $\bar{x}_2$, the critical value of $\bar{x}$ at the 1 per cent significance level, the formulae below are used:

$$\bar{x}_1 = \mu + t_{0.05} \times \frac{s}{\sqrt{n}}$$

$$\bar{x}_2 = \mu + t_{0.01} \times \frac{s}{\sqrt{n}}$$

The corresponding decision rules are given below.

1. Whenever $\bar{x}_1 < \bar{x} < \bar{x}_2$ the system is on the alert for an increase in μ.
2. Whenever $\bar{x}_2 < \bar{x}$ it is almost certain that an increase in μ has taken place.

When the test is for a decrease in the population mean the decision rules below are applied.

1. Whenever $\mu - t_{0.05} \times \dfrac{s}{\sqrt{n}} > \bar{x} > \mu - t_{0.01} \times \dfrac{s}{\sqrt{n}}$ the system is on the alert for a decrease in μ.

2. Whenever $\mu - t_{0.01} \times \dfrac{s}{\sqrt{n}} > \bar{x}$ it is almost certain that a decrease in μ has taken place.

When the test is for a change in the population mean the decision rules below are applied.

1. Whenever $\bar{x}$ lies between $\mu + t_{0.025} \times \dfrac{s}{\sqrt{n}}$ and $\mu + t_{0.005} \times \dfrac{s}{\sqrt{n}}$ or between $\mu - t_{0.025} \times \dfrac{s}{\sqrt{n}}$ and $\mu - t_{0.005} \times \dfrac{s}{\sqrt{n}}$ the system is on the alert for a possible change in μ.

2. Whenever $\bar{x}$ is greater than $\mu + t_{0.005} \times \dfrac{s}{\sqrt{n}}$ or less than $\mu - t_{0.005} \times \dfrac{s}{\sqrt{n}}$ it is almost certain that a change in μ has taken place.

Exercise 6.2

Determine critical values at the 5 per cent and 1 per cent levels in the following examples.

1. $\mu = 401.8$, $\sigma = 3.95$, $n = 8$, for an increase in μ.
2. $\mu = 50.9$, $\sigma = 5.56$, $n = 10$, for an increase in μ.
3. $\mu = 126.7$, $\sigma = 29.4$, $n = 15$, for a change in μ.
4. $\mu = 456.8$, $n = 18$, $s = 28.5$, for a change in μ.
5. $\mu = 79.3$, $n = 26$, $s = 4.62$, for a decrease in μ.
6. $\mu = 3025$ mm, $n = 20$, $\sigma = 210.5$ mm, for a decrease in μ.
*7. A machine produces plastic blocks with mean mass 500 g and standard deviation 1 g. To test for an increase in the mean mass using samples of 20 calculate critical values of $\bar{x}$ at the 5 per cent and 1 per cent significance levels. Determine corresponding decision rules.
*8. With the data of question (7) determine critical values and decision rules for testing a change in the mass of the population mean using samples of 20.
*9. With the population data of question (6) in Exercise 6.1 determine decision rules for samples of 25 to determine:
   (a) whether or not an increase in the mean breaking strength has taken place;
   (b) whether or not a decrease in the mean breaking strength has taken place;
   (c) whether or not a change in the mean breaking strength has taken place.
*10. A machine is set up to produce nails with a mean length of 3.000 cm. Determine decision rules for the mean lengths of samples of 12 nails which have a standard deviation of 0.3 cm which will test for:
   (a) an increase in the mean length of the population;
   (b) a decrease in the mean length of the population;
   (c) a change in the mean length of the population.

6.3 Type I and type II errors

When decision rules are applied to the problems of rejecting defects in a manufacturing process and to the acceptance or rejection of hypotheses, two types of error may arise.

A type I error is one when a product which is not defective is rejected and when a hypothesis which is correct is not accepted.

A type II error is one when a product which is defective is not rejected and when a hypothesis which is false is not rejected, or when it is accepted.

There is no system of decision rules which can give absolute certainty that neither of these errors will occur. However, the decision rules discussed do minimize the probabilities of such errors occurring.

To help us to remember the distinction between the two types of error Table 6.1 often proves useful.

Table 6.1 *Type I and type II errors*

Hypothesis	Decision	
	Accept	Reject
True	Correct	Type I error
False	Type II error	Correct

6.4 Paired sample tests

When two identical samples are taken from a large population and subjected to the same observations in two different circumstances, the sample is said to be a paired sample. The following are examples of paired samples:

1. A group of people treated with two different drugs to test the relative effectiveness of the drugs.
2. A set of surfaces treated with two different kinds of paint to test how the paints stand up to certain climatic conditions.
3. The effect of coaching or non-coaching on a set of candidates for a particular test.
4. The comparison of two methods of hardening a set of metals.

Example

Twenty rooms consisting of ten different kinds of identical pairs are selected. Two different kinds of heat insulation systems are tested: system *A* and system *B*. One group of ten rooms is treated with system *A* and the other group of ten is treated with system *B*. The rooms are tested individually for average heat loss at a standard difference of outer and inner temperatures both in the untreated and the treated state, and the percentage savings in heat are determined. From the data below determine whether or not any particular system of heat insulation produces a higher mean percentage saving.

Room numbers:	1	2	3	4	5	6	7	8	9	10
Percentage saving system *A*:	35.0	27.3	23.8	33.6	44.1	42.7	36.4	30.1	35.0	29.4
Percentage saving system *B*:	32.2	32.2	32.9	35.7	52.5	35.0	44.8	39.9	41.3	21.7

Represent the differences between systems *A* and *B* by the symbol x. They are tabulated below.

x: 2.8 -4.9 -9.1 -2.1 -8.4 7.7 -8.4 -9.8 -6.3 7.7

Take the null hypothesis that there is no significant difference between system *A* and system *B*, i.e. the mean difference, $\bar{x}$, does not differ significantly from $\mu = 0$. From the data:

$$\bar{x} = -3.08$$

Then $s = 6.8283396$

A decision has to be made about the alternative hypothesis and the nature of the test to be applied. Is it to be a one-tailed test or a two-tailed test? In this case, since $\bar{x}$ is negative, it would seem that we ought to be testing whether or not system *B* is a better system than system *A*, in other words, whether or not $\mu < 0$. That involves a one-tailed test. Consequently we take:

$$H_0 : \mu = 0$$
$$\text{and} \quad H_1 : \mu < 0$$

For nine degrees of freedom, with a one-tailed test:

$$t_{0.05} = 1.833$$
$$\text{and} \quad t_{0.01} = 2.821$$

The critical values are:

$$\bar{x}_1 = 0 - 1.833 \times \frac{6.8283396}{\sqrt{10}}$$
$$= -3.9580163$$
$$\approx -3.96$$

$$\text{and} \quad \bar{x}_2 = 0 - 2.82 \times \frac{6.8283396}{\sqrt{10}}$$
$$= -6.0892558 \approx -6.09$$

In this case:

$$|\bar{x}| < |\bar{x}_1|$$
$$\text{Since} \quad 3.08 < 3.96$$

Consequently, even at the 5 per cent significance level, the evidence does not support a rejection of H_0. We conclude that there is no evidence here for believing that system B produces a higher mean percentage heat saving than does system A.

Exercise 6.3

Using the data in the following examples compare the means of the paired samples.

1. | Item: | 1 | 2 | 3 | 4 | 5 | 6 |
   |---|---|---|---|---|---|---|
   | A: | 42 | 32 | 43 | 44 | 33 | 43 |
   | B: | 36 | 30 | 43 | 43 | 37 | 40 |

2. | Item: | 1 | 2 | 3 | 4 | 5 | 6 | 7 |
   |---|---|---|---|---|---|---|---|
   | P: | 24 | 36 | 55 | 53 | 41 | 27 | 53 |
   | Q: | 25 | 39 | 67 | 75 | 57 | 51 | 72 |

3. | Item: | 1 | 2 | 3 | 4 | 5 | 6 | 7 | 8 | 9 | 10 |
   |---|---|---|---|---|---|---|---|---|---|---|
   | X: | 200 | 240 | 218 | 228.8 | 232 | 235 | 192 | 188 | 210 | 214 |
   | Y: | 195 | 202 | 214 | 190 | 179 | 168 | 138 | 191 | 173 | 134 |

4. Ten amateur snooker players were given the same practical test. From the same position of the balls on the table their initial breaks from that position were noted. Each player was given similar coaching and tested again in the same way as before. The data are given in the table below.

Player:	1	2	3	4	5	6	7	8	9	10
Break before coaching:	15	22	23	9	3	18	27	8	44	23
Break after coaching:	22	20	29	13	23	34	23	10	56	38

   Does the evidence indicate that coaching did have an appreciable effect on the performance of the players?

7

Basic Ideas of Correlation

7.1 Scatter diagrams

Bivariate data are observations or measurements of two variable quantities arranged in a form, usually tabular, so that corresponding values of the two variables can easily be identified. Where the variables are referred to as x and y it is customary to identify x with the independent variable and y with the dependent variable. This is preliminary to the search for a possible relationship between the variables.

The data may be collected from an experiment, from some economic survey, from observation of some natural source, or from some other origin. The data are then examined statistically to discover whether there is any evidence of a relationship between the variables. The statistical approach is often the initial step in uncovering a connection between two occurrences. For instance, the connection between smoking and the incidence of lung cancer rests mainly on statistical evidence.

An initial step in the search for a relationship is the plotting of a scatter diagram. Fig. 7.1 represents such a diagram. It relates the average wage per head of the population to the number of new cars sold per year. The points on the diagram are not accurately plotted points based on accurate information but represent a sketch of what such material might reveal. Note that there is a definite trend of the data around a line which is indicated in the diagram. When a scatter diagram does show such a trend of the data, that is taken as evidence that there is a likelihood of a linear relationship between the variables.

Fig. 7.2 is a sketch of a scatter diagram showing the relationship between

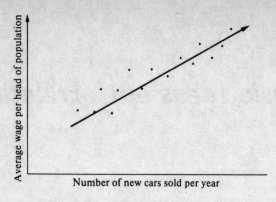

Figure 7.1

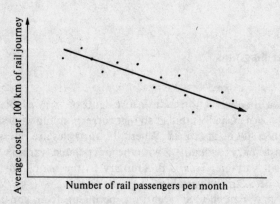

Figure 7.2

the average cost per 100 km of a rail journey and the number of rail passengers per month. Again there seems to be evidence of a linear trend. This time the line has quite a different direction from that in Fig. 7.1.

Fig. 7.3 is a scatter diagram showing a different characteristic.

Fig. 7.1 appears to indicate some form of relationship between the average wage per head of the population and the number of cars sold because all the dots lie close to the line indicated in the diagram and, as one variable increases, so does the other. The gradient of the line in Fig. 7.1 is positive.

Fig. 7.2 appears to indicate another form of relationship between the average cost per 100 km of a rail journey and the number of rail passengers per month. As one variable increases the other decreases. The gradient of this line is negative.

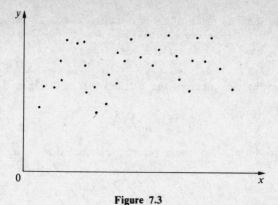

Figure 7.3

Fig. 7.3 appears to indicate no relationship between the two variables y and x.

When the points on a scatter diagram lie close to a line then there is a strong indication of a linear relationship between the two variables. In that event there is said to be a correlation between the two variables. When the correlation is positive that indicates that an increase in the independent variable produces an increase in the dependent variable and vice versa. In that case the line to which the scatter of points approximates has a positive gradient.

When the correlation is negative that indicates that an increase in one variable produces a decrease in the other. In that event the line to which the scatter of points approximates has a negative gradient.

One method of arriving at a measure of this correlation is to determine the equation of the line of best fit for the data. That method is discussed in Chapter 8. An alternative method is to determine a value which may be attached to the correlation. It is called the *correlation coefficient* or the *product moment coefficient of correlation*.

7.2 The product moment correlation coefficient

Suppose that we wish to determine the product moment correlation coefficient between two variables x and y for which we possess n pairs of corresponding values, x_r, y_r, when r takes the values 1 to n inclusive.

Suppose also that the mean of the n values of x is $\bar{x}$. Therefore:

$$\bar{x} = \frac{1}{n} \sum_{r=1}^{n} x_r$$

Suppose also that the mean of the n values of y is $\bar{y}$. Therefore:

$$\bar{y} = \frac{1}{n} \sum_{r=1}^{n} y_r$$

The standard deviation of the x values is calculated to be σ_x. Therefore:

$$\sigma_x = \sqrt{\frac{1}{n} \sum_{r=1}^{n} (x_r - \bar{x})^2}$$

which is sometimes written as:

$$\sqrt{\frac{1}{n} \sum (x - \bar{x})^2}$$

$$\text{or} \quad \sqrt{\frac{1}{n} \sum x^2 - \bar{x}^2}$$

The standard deviation of the y values is calculated to be σ_y. Therefore:

$$\sigma_y = \sqrt{\frac{1}{n} \sum_{r=1}^{n} (y_r - \bar{y})^2}$$

which is sometimes written as:

$$\sqrt{\frac{1}{n} \sum (y - \bar{y})^2} \quad \text{or} \quad \sqrt{\frac{1}{n} \sum y^2 - \bar{y}^2}$$

The correlation coefficient, r, is given by the formula:

$$r = \frac{\sum (x - \bar{x})(y - \bar{y})}{n . \sigma_x . \sigma_y} \qquad 7.1$$

The expression $\frac{1}{n} \sum (x - \bar{x})(y - \bar{y})$ is called the covariance of x and $y (\sigma xy)$.

Therefore an alternative method of defining r is by the formula:

$$r = \frac{\text{covariance of } x \text{ and } y}{\text{standard deviation of } x \times \text{standard deviation of } y} \qquad 7.2$$

In symbol form 7.2 becomes:

$$r = \frac{\sigma_{xy}}{\sigma_x . \sigma_y} \qquad 7.3$$

An alternative version of *7.3* is the formula:

$$r = \frac{\sum (x - \bar{x})(y - \bar{y})}{\sqrt{\sum (x - \bar{x})^2} \cdot \sqrt{\sum (y - \bar{y})^2}}$$

7.4

Examples

Calculate the product moment correlation coefficients for the data in the examples below.

1.

x:	1	2	3	4	5	6	7
y:	3	6	9	12	15	18	21

From the data, $\bar{x} = 4$, $\bar{y} = 12$, $n = 7$, σ_x, also sometimes written $S_x = 2$, and σ_y, also sometimes written $S_y = 6$. Then:

$$\frac{1}{n} \sum (x - \bar{x})(y - \bar{y}) = S_{xy} = \sigma_{xy} = 12$$

Therefore:

$$r = \frac{12}{2 \times 6} = 1$$

In this example it does not require a calculation to show that there is a perfect linear relationship between the variables y and x. In fact $y = 3x$. This is a line through 0 with gradient 3. The value $r = 1$ is in fact the highest positive value that r may take. It demonstrates a complete positive correlation between the variables: an increase in one produces a perfectly linearly related increase in the other.

2.

x:	1	2	3	4	5	6	7	8	9
y:	36	32	28	24	20	16	12	8	4

From the data, $\bar{x} = 5$, $\bar{y} = 20$, $n = 9$, $\sigma_x = S_x = 2.5819889$, and $\sigma_y = S_y = 10.327956$. Then:

$$\sigma_{xy} = S_{xy} = \frac{1}{n} \sum (x - \bar{x})(y - \bar{y}) = -\frac{80}{3}$$

Therefore:

$$r = -\frac{80}{3 \times 2.5819889 \times 10.327956} = -1$$

Again, simple observation of the data shows that there is a linear relation between the variables. Here $y = 40 - 4x$. This is a line with gradient -4 and intercept 40. The value $r = -1$ is the lowest negative value which r may take. It demonstrates a perfect negative correlation between the variables: an increase in one variable produces a perfect linearly related decrease in the other. No correlation coefficient between pairs of corresponding variables lies outside the range -1 to $+1$.

3.
$$x: \quad 2 \quad 4 \quad 6 \quad 8 \quad 10 \quad 12$$
$$y: \quad 3 \quad 6 \quad 9 \quad 9 \quad 6 \quad 3$$

From the data, $\bar{x} = 7$, $\bar{y} = 6$, $n = 6$, $\sigma_x = S_x = 3.4156503$, and $\sigma_y = S_y = 2.4494897$. Then:

$$\frac{1}{n}\sum(x - \bar{x})(y - \bar{y}) = \sigma_{xy} = S_{xy} = 0$$

Therefore $r = 0$.
In this example there is no correlation at all. A glance at the data shows that the relative behaviour of the first three pairs of corresponding values is opposite to that of the second three pairs of corresponding values.

Exercise 7.1

Draw scatter diagrams and calculate the correlation coefficients for the following sets of data.

1.
$$x: \quad 1 \quad 2 \quad 3 \quad 4 \quad 5 \quad 6 \quad 7 \quad 8 \quad 9 \quad 10$$
$$y: \quad 4 \quad 6 \quad 12 \quad 12 \quad 25 \quad 24 \quad 28 \quad 25 \quad 36 \quad 29$$

2.
$$p: \quad 19 \quad 21 \quad 16 \quad 18 \quad 17 \quad 14 \quad 13 \quad 12 \quad 10 \quad 11$$
$$q: \quad 28 \quad 26 \quad 24 \quad 22 \quad 18 \quad 17 \quad 16 \quad 14 \quad 13 \quad 12$$

3.
$$U: \quad 51 \quad 55 \quad 59 \quad 73 \quad 63 \quad 62 \quad 51 \quad 59 \quad 58 \quad 89$$
$$V: \quad 31 \quad 40 \quad 62 \quad 60 \quad 61 \quad 62 \quad 62 \quad 57 \quad 51 \quad 63$$

4.
$$x: \quad 62 \quad 70 \quad 74 \quad 62 \quad 69 \quad 100 \quad 70 \quad 73$$
$$y: \quad 10.4 \quad 10.6 \quad 10.2 \quad 10.7 \quad 10.2 \quad 10.0 \quad 10.3 \quad 10.4$$

5.
$$x: \quad 41.6 \quad 40.1 \quad 39.6 \quad 38.0 \quad 37.2 \quad 36.8 \quad 36.4$$
$$y: \quad 1.65 \quad 1.77 \quad 1.79 \quad 1.44 \quad 1.42 \quad 1.64 \quad 1.66$$
$$x: \quad 35.2 \quad 32.1 \quad 30.4 \quad 30.2 \quad 31.5 \quad 34.2 \quad 36.3$$
$$y: \quad 1.48 \quad 1.20 \quad 1.31 \quad 1.22 \quad 1.39 \quad 1.28 \quad 1.41$$

6. A company's monthly sales (S) are tabulated against its advertising

expenditures (E) for the previous month. Calculate the correlation coefficient.

Sales (S)($£10^3$):	550	620	304	134	498	360
Advertising expenditures (E)($£10^3$):	35.6	45.2	68.2	60.8	72.8	76.2
S:	138	438	434	192	372	474
E:	75.4	58.2	21.8	28.0	17.0	21.8

7. A variable speed machine operates a drill head so that the speed of the cutting tool can be varied. The data below relate life of drill head in hours to cutting speed in metres per second. Calculate the correlation coefficient.

Cutting speed (s) (m/s):	1.5	1.8	2.3	2.9	3.8
Life of drill head (l) (hours):	147	150	125	91	88
Cutting speed (s):	4.0	3.7	2.8	2.0	
Life of drill head (l):	54	80	110	113	

7.3 Significance of the correlation coefficient

When two sets of data are being compared to detect whether or not there is a relationship between them we need to know what values of r are significant. For instance, in a given investigation, do we interpret that a value of r = 0.87 does show a definite relationship between the two sets of data? On the face of it such a value would appear to indicate that there was a definite connection. However, before we proceed to determine significance limits and decision rules, it must be said that even when the numerical values of the two sets of data produce a high numerical value of r it might be ridiculous to infer a relation between them. For instance, it might be possible to obtain data for the maximum temperatures in Bradford on certain random dates and data for the sales of sugar in Hong Kong on the same dates which produce a high value of r, and it seems highly unlikely that one would affect the other. So the nature of the data must be examined carefully before a calculation of r is attempted.

On the other hand, although on the surface a relationship between two variables may be obscure, that should not prohibit a determination of r. There may, in fact, be some hidden connection which escapes a superficial investigation.

The coefficient of determination

The value r^2 measures the proportion of total variation of y which can be explained by variations in x. When $r = 0.87$, $r^2 = 0.7569 \approx 0.76$, and that means that 76 per cent of the variation in y can be explained by variations in x. The remaining 24 per cent of the variation in y is explained by other factors.

In fact the graph relating r^2 to r is parabolic. When r is doubled r^2 is quadrupled. When r is trebled r^2 is multiplied by nine. The value of r^2, therefore, might be used as an indicator of the degree of relationship between the two sets of data.

The significance of r

A number of different formulae have been devised to test the significance of r. When the sample is drawn from an uncorrelated, normal population the t-distribution is used and the following formula applied:

$$t = \sqrt{\frac{r^2(n-2)}{(1-r^2)}} \qquad 7.5$$

Here r is taken to be a *t-distribution* with $(n-2)$ *degrees of freedom* and t-tables for a *two-tailed* test are applied.

In the outlines of the methods described below the total populations from which the two samples have been drawn are taken to have a correlation coefficient of c. Two methods may be applied.

The first is to rearrange the formula for t to determine r by squaring 7.5:

$$t^2 = \frac{r^2(n-2)}{(1-r^2)}$$

i.e. $\qquad t^2(1-r^2) = r^2(n-2)$

$$t^2 = r^2(n-2+t^2)$$

$$r^2 = \frac{t^2}{(n-2+t^2)}$$

i.e. $\qquad r = \frac{t}{\sqrt{n-2+t^2}} \qquad 7.6$

For sample size n, with $n-2$ degrees of freedom determine t_1, the two-tailed value of t at the 5 per cent significance level, and t_2, the two-tailed value of t

at the 1 per cent significance level. Then, from 7.6:

$$r_1 = \frac{t_1}{\sqrt{n-2+t_1^2}}$$

$$r_2 = \frac{t_2}{\sqrt{n-2+t_2^2}}$$

Now take the null hypothesis that the populations are uncorrelated and the alternative hypothesis that they are correlated:

$$H_0: c = 0$$
$$H_1: c \neq 0$$

The decision rules are:

1. Accept H_0 and reject H_1 if $r < r_1$.
2. Do not reject H_0 and do not accept H_1 if $r_1 < r < r_2$.　　　7.7
3. Reject H_0 and accept H_1 if $r > r_2$.

Table 7.1

n	r_1	r_2	t_1	t_2
10	0.6318962	0.764558	2.306	3.355
11	0.6020421	0.7348034	2.262	3.250
12	0.575959	0.7078572	2.228	3.169
13	0.5529452	0.6835503	2.201	3.106
14	0.5324456	0.6614317	2.179	3.055
15	0.5139129	0.6411102	2.160	3.012
16	0.4973463	0.6226109	2.145	2.977
17	0.482068	0.6055433	2.131	2.947
18	0.4682936	0.5897432	2.120	2.921
19	0.4555621	0.5750362	2.110	2.898
20	0.4437766	0.5613766	2.101	2.878
21	0.4328535	0.5487198	2.093	2.861
22	0.4227196	0.536754	2.086	2.845
23	0.4133107	0.5255716	2.080	2.831
24	0.404407	0.5151339	2.074	2.819
25	0.396125	0.5051358	2.069	2.807
26	0.3882602	0.4958159	2.064	2.797
27	0.3809358	0.4868736	2.060	2.787
28	0.3739595	0.4785491	2.056	2.779
29	0.3673039	0.470551	2.052	2.771
30	0.3609445	0.4628578	2.048	2.763

Table 7.1 gives a list of values of r_1 and r_2 related to sample sizes, n, calculated from formula 7.6 using the appropriate values of t, which are also given in the table. The values of r_1 and r_2 have been listed in the table as they

were obtained by calculator. They may be reduced to whatever degree of accuracy is required.

Examples

1. When $n = 20$ and $r = 0.592$ then $r > r_2$ (0.5613766). Consequently, at the 1 per cent significance level, we would accept the alternative hypothesis that there was correlation.
2. When $n = 15$ and $r = 0.581$ then $r_1 < r < r_2$. *Either* we would not reject H_0 and not accept H_1 *or* we might be prepared to accept H_1 at the 5 per cent significance level.
3. When $n = 25$ and $r = 0.382$ then $r < r_1$. Consequently we would accept H_0.

The second method is, when r has been calculated from the data, to use formula 7.5 to determine the corresponding value of t. Suppose that for sample size n, i.e. $n - 2$ degrees of freedom, t_1 is the two-tailed value of t at the 5 per cent significance level and t_2 is the two-tailed value of t at the 1 per cent significance level. With this approach alternative decision rules are those given below:

1. Accept H_0 and reject H_1 if $t < t_1$.
2. Do not reject H_0 and do not accept H_1 if $t_1 < t < t_2$.
3. Reject H_0 and accept H_1 if $t > t_2$.

Example

With the data of question (1) in Exercise 7.1 determine whether there is any significant correlation between y and x. From the data, $\bar{x} = 5.5$, $\bar{y} = 20.1$, $n = 10$, $\sigma_x = 2.8722813$ and $\sigma_y = 10.231813$. The values of $x - \bar{x}$ and $y - \bar{y}$ are tabulated below:

$x - \bar{x}$:	-4.5	-3.5	-2.5	-1.5	-0.5	0.5	1.5
$y - \bar{y}$:	-16.1	-14.1	-8.1	-8.1	4.9	3.9	7.9
$x - \bar{x}$:	2.5	3.5	4.5				
$y - \bar{y}$:	4.9	15.9	8.9				

Therefore:

$$\frac{1}{n} \Sigma (x - \bar{x})(y - \bar{y}) = \sigma_{xy} = 27.35$$

These values give:

$$r = \frac{27.35}{2.8722813 \times 10.231813}$$

$$= 0.9306315$$

$$\approx 0.931$$

Take the null hypothesis to be that there is no correlation between y and x and the alternative hypothesis to be that there is a correlation. That is:

$$H_0: c = 0$$
$$H_1: c \neq 0$$

From Table 7.1:

$$r_1 \approx 0.632 \text{ and } r_2 \approx 0.765$$

Therefore, in this instance, $r > r_2$. Consequently accept H_1 at the 1 per cent significance level. There is real evidence of a correlation between y and x. *Alternatively*, when $r = 0.9306315$, by formula *7.5*:

$$t = \sqrt{\frac{r^2(n-2)}{(1-r^2)}}$$

$$= 7.1926992$$

$$\approx 7.193$$

From t-tables, with eight degrees of freedom:

$$t_1 = 2.306$$
$$t_2 = 3.355$$

Therefore, in this instance, $t > t_2$. Consequently, at the 1 per cent significance level, accept H_1. There is real evidence of correlation between y and x.

Exercise 7.2

1. Using t-distribution values calculate values of r_1 and r_2 for values of n from 5 to 9 inclusive.

   Take questions (2) to (7) inclusive to be based on questions (2) to (7) of Exercise 7.1. With the data in each of those questions determine whether or not there is real evidence of correlation between the two sets of data.

8. The table below gives the annual rainfall and annual sunshine figures for a particular seaside resort in the UK in corresponding years.

Rainfall (R) (cm):   320   378   384   409   394   335
Sunshine (S) (h):   1894 1855 1712 1940 1841 1896
Rainfall (R) (cm):   452   348   442   422   401
Sunshine (S) (h):   1561 1963 1580 1856 1654

Determine whether or not there is any significant correlation between the annual hours of sunshine and the annual rainfall at that resort.

8
Basic Ideas of Regression

8.1 Dependent and independent variables

Equations such as:

$$y = mx + c \qquad \text{(i)}$$

$$y = ax^2 + bx + c \qquad \text{(ii)}$$

$$y = 5 \sin 2x \qquad \text{(iii)}$$

$$y = \frac{x}{4}e^{-x} \qquad \text{(iv)}$$

express a relationship between a quantity, x, which may vary, and another quantity, y, which also varies. Both x and y are called variables. The variation in y depends on the variation in x in a manner determined by the nature of the function on the right of the equation. Therefore x is called the independent variable and y is called the dependent variable.

Only equation (i) above represents a linear relation between the two variables. When there is perfect correlation between the data for two variables the relationship is a linear one. That is, when r takes either of the values $+1$ or -1, y is linearly related to x.

When r takes a value numerically less than 1 there is a certain degree of linear relation between the two sets of data unless r is equal to zero. In the previous chapter we concentrated on determining values of r and on deciding how significant those values were in deciding correlation. In this chapter more attention will be paid to the graphical representation of the data and its use in determining the most appropriate linear connection between the variables.

113

8.2   Scatter diagrams

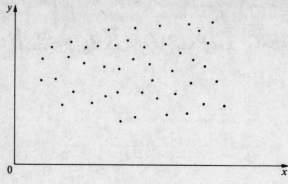

Figure 8.1

Suppose the data from two sets of observations or measurements, x and y, can be represented in a diagram such as Fig. 8.1. We wish to know whether or not there is a linear relation connecting x and y using a method different from the correlation coefficient method of the last chapter. The method of deciding the line which seems to best fit the data using the eye alone as a judge is far from satisfactory. The basic ideas on which the more refined method rests are given below.

Method 1

Fig. 8.2 represents a scatter diagram in which vertical strips of equal width are drawn. The strips divide the plotted points into groups. Within each

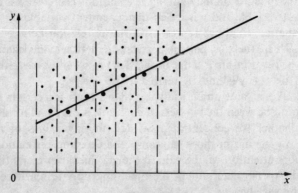

Figure 8.2

strip the mean of the group in that strip is determined and marked with a cross. A line which is the best fit for the mean points is then determined. That line is the line of regression of y on x. The equation of the line is denoted by:

$$y = mx + c \qquad\qquad 8.1$$

Method 2

In Fig. 8.3 the plotted points are divided into different groups by horizontal strips of equal width. Again the mean of each group in a strip is marked with a cross and the line of best fit for the mean points is determined. That line is the line of regression of x on y. The equation of the line is denoted by:

$$x = m'y + c' \qquad\qquad 8.2$$

$$\text{i.e.} \quad y = \frac{1}{m'}x - c' \qquad\qquad 8.3$$

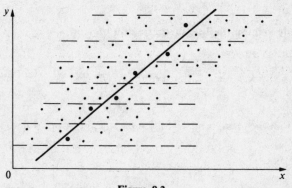

Figure 8.3

In most cases the two regression lines are not identical. When they are, then:

$$m = \frac{1}{m'}$$

$$\text{i.e.} \quad mm' = 1 \qquad\qquad 8.4$$

When that happens there is an exact linear relationship between y and x. When either m or m' = 0, then:

$$mm' = 0 \qquad\qquad 8.5$$

In that event there is no correlation between y and x. The value of the product mm' plays an important role in the decision about whether or not there is a close correlation between the two variables. This is because:

$$r^2 = mm' \qquad\qquad 8.6$$

where r is the coefficient of correlation. When m and m' are both positive r is taken to be positive. When m and m' are both negative r is taken to be negative. The lines of regression both pass through the mean point of all the plotted points: they pass through $(\bar{x}, \bar{y})$. Therefore the regression line of y on x may be written:

$$y - \bar{y} = m(x - \bar{x}) \qquad\qquad 8.7$$

The line of regression of x on y may be written:

$$x - \bar{x} = m'(y - \bar{y}) \qquad\qquad 8.8$$

The methods which follow outline the exact procedure adopted in determining the lines of regression even though the results of the analysis are assumed to be true.

8.3   To determine the line of regression of y on x

In Fig. 8.4 suppose the plotted points are $P_1, P_2, P_3 \ldots P_n$ where:

$$P_r \equiv (x_r, y_r)$$

Suppose that the mean point of all the Ps is $(\bar{x}, \bar{y})$. Suppose also that the line

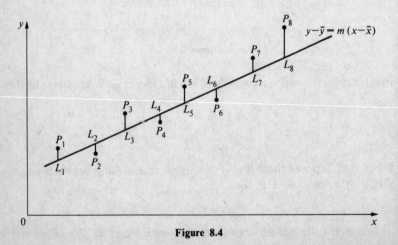

Figure 8.4

of best fit for a line of regression of y on x is:

$$y - \bar{y} = m(x - \bar{x})$$

Lines $P_1L_1, P_2L_2, P_3L_3, \ldots P_rL_r \ldots P_nL_n$ are drawn parallel to Oy to meet the line of regression of y on x in $L_1, L_2, L_3 \ldots L_r \ldots L_n$. Then the line of regression is defined to be the line such that:

$$(P_1L_1)^2 + (P_2L_2)^2 + (P_3L_3)^2 + \ldots + (P_nL_n)^2$$

is a minimum. It will be assumed that this happens when m has the value:

$$\frac{r\sigma_y}{\sigma_x} \qquad \text{8.9}$$

Therefore the equation of the line of regression of y on x may be expressed:

$$\left(\frac{y - \bar{y}}{\sigma_y}\right) = r\left(\frac{x - \bar{x}}{\sigma_x}\right) \qquad \text{8.10}$$

Therefore

$$m = r\frac{\sigma_y}{\sigma_x}$$

$$= \frac{\sigma_{xy}}{\sigma_x \cdot \sigma_y} \cdot \frac{\sigma_y}{\sigma_x} \qquad \text{8.11}$$

$$= \frac{\sigma_{xy}}{\sigma_x^2} \qquad \text{8.12}$$

$$= \frac{\sum (x - \bar{x})(y - \bar{y})}{n \cdot \sigma_x^2} \qquad \text{8.13}$$

8.4 To determine the line of regression of x on y

In Fig. 8.5 suppose the line of best fit for a regression line of x on y is:

$$x - \bar{x} = m'(y - \bar{y})$$

where $(\bar{x}, \bar{y})$ is the mean of all the plotted points. Draw lines $P_1M_1, P_2M_2, P_3M_3 \ldots P_rM_r \ldots P_nM_n$ parallel to Ox to meet the lines of regression in $M_1, M_2, M_3 \ldots M_r \ldots M_n$ respectively. The line of regression of x on y is defined to be the line such that:

$$(P_1M_1)^2 + (P_2M_2)^2 + (P_3M_3)^2 + \ldots + (P_rM_r)^2 + \ldots + (P_nM_n)^2$$

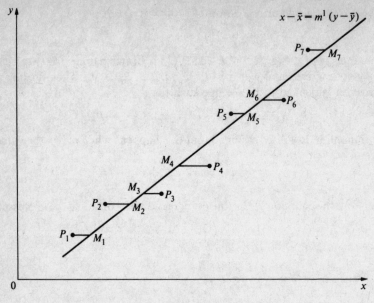

Figure 8.5

is a minimum. It will be assumed that this happens when m' has the value:

$$r \cdot \frac{\sigma_x}{\sigma_y} \qquad 8.14$$

The line of regression of x on y is the line with the equation:

$$\left(\frac{x - \bar{x}}{\sigma_x}\right) = r\left(\frac{y - \bar{y}}{\sigma_y}\right) \qquad 8.15$$

Therefore:

$$m' = r\frac{\sigma_x}{\sigma_y}$$

$$= \frac{\sigma_{xy}}{\sigma_x \cdot \sigma_y} \cdot \frac{\sigma_x}{\sigma_y}$$

$$= \frac{\sigma_{xy}}{\sigma_y^2} \qquad 8.16$$

$$= \frac{\Sigma(x - \bar{x})(y - \bar{y})}{n \cdot \sigma_y^2} \qquad 8.17$$

8.5   Regression lines

Examples

1. Plot the data in the table below and determine whether a linear relation exists between the two variables.

$$x:\quad 2\quad 4\quad 6\quad 8\quad 10$$
$$y:\quad 2.2\quad 3.4\quad 4.3\quad 5.5\quad 6.6$$

From the data, $\bar{x} = 6$, $\bar{y} = 4.4$.

$$x - \bar{x}:\quad -4\quad -2\quad 0\quad 2\quad 4$$
$$y - \bar{y}:\quad -2.2\quad -1\quad -0.1\quad 1.1\quad 2.2$$

$$\sum(x - \bar{x})(y - \bar{y}):\ 8.8 + 2 + 2.2 + 8.8 = 21.8$$

$$\sigma_{xy} = \frac{1}{n}\sum(x - \bar{x})(y - \bar{y}) = \frac{1}{5} \times 21.8 = 4.36$$

$$\sigma_x^2 = \frac{1}{5}[(-4)^2 + (-2)^2 + 0 + 2^2 + 4^2] = \frac{1}{5} \times 40 = 8$$

$$\sigma_y^2 = \frac{1}{5}[(-2.2)^2 + (-1)^2 + (-0.1)^2 + 1.1^2 + 2.2^2]$$

$$= \frac{1}{5} \times 11.9 = 2.38$$

Therefore:

$$m = \frac{4.36}{8} = 0.545$$

$$c = \bar{y} - m\bar{x} = 4.4 - 0.545 \times 6 = 1.13$$

The line of regression of y on x is:

$$y = 0.545x + 1.13 \tag{i}$$

Again:

$$m' = \frac{4.36}{2.38} = 1.8319328$$

$$c' = \bar{x} - m'\bar{y}$$

$$= 6 - 4.4 \times 1.8319328$$

$$= -2.0605042$$

The line of regression of x on y is:

$$x = 1.8319328y - 2.0605042 \qquad \text{(ii)}$$

More approximately, (ii) is:

$$x = 1.832y - 2.061$$

The value of $mm' = 0.9984034$. This is close to 1. Therefore there is a high degree of correlation between y and x. Fig. 8.6 represents the plotted points and the two lines of regression.

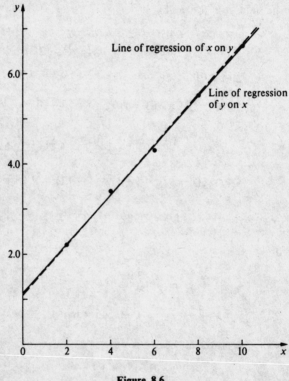

Figure 8.6

2. Draw lines of regression for the data in question (7) of Exercise 7.1. Call the variables for cutting speed x and those for the life of the drill head y.

x:	1.5	1.8	2.3	2.9	3.8	4.0	3.7	2.8	2.0
y:	147	150	125	91	88	54	80	110	113

From the data:

$$\bar{x} = 2.7555555, \ \bar{y} = 106.44444$$
$$m = -31.884365$$
$$c = 194.30358$$
$$\sigma_x^2 = 0.7580247$$
$$\sigma_y^2 = 885.58025$$
$$r = -0.9328364$$
$$m' = -0.0292569$$
$$c' = \bar{x} - m'\bar{y} = 5.8697856$$

The line of regression of y on x is:

$$y = -31.884365x + 194.30358$$

which approximates to:

$$y = -31.88x + 194.3$$

The line of regression of x on y is:

$$x = -0.0292569y + 5.8697856$$

which approximates to:

$$x = -0.029y + 5.870$$

Fig. 8.7 represents the plotted data and the two regression lines. Again there is a high degree of correlation. In this instance, as the cutting speed increases, the life of the drill head decreases.

Exercise 8.1

For questions (1) to (6) inclusive of this exercise take the data from questions (1) to (6) of Exercise 7.1. In each case plot the points representing the data and determine the equations of both lines of regression. Mark in these lines on the graphs.

8.6 The interpolation and extrapolation of additional data from the equations of the lines of regression

Additional data may be obtained about the dependent and independent variables either graphically from the graphs of the lines of regression when

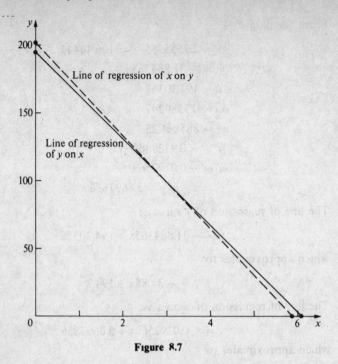

Figure 8.7

they have been drawn or algebraically from the equations of the lines of regression. At earlier levels much attention was paid to graphical methods of interpolation and extrapolation. Consequently, in this context, we shall concentrate on the use of the equations to obtain additional data.

The equation of the line of regression of y on x may be used to determine values of y for given values of x. When the value of x lies within the range of values of x in the data the process is called interpolation. When the value of x lies outside the range of values of x the process is called extrapolation.

The use of the equation of the line of regression to determine a value of y corresponding to a value of x really amounts to a prediction of what y will be for a given value of x. This assumes that the trend, represented by the line of regression, is a constant trend and can be forecasted. In practice, however, all sorts of influences are brought to bear to modify the general trend. Therefore we must always be on our guard for errors in predictions. The predictions of extrapolation are subject to even greater error than those of interpolation. The examples below illustrate the procedures for interpolation and extrapolation.

Examples

1. From the data of Example (1) in section 8.5 determine values of y when x takes the values 5, 12 and 1 and the values of x when y takes the values 2.6, 3.8, 4.9 and 8.2.

    The line of regression of y on x is $y = 0.545x + 1.13$. Therefore:

when $x = 5$	$y = 3.855$
when $x = 12$	$y = 7.67$
when $x = 1$	$y = 1.675$

    Further:

when $x = 2$	$y = 2.22$
when $x = 4$	$y = 3.31$
when $x = 6$	$y = 4.4$
when $x = 8$	$y = 5.49$
when $x = 10$	$y = 6.58$

    The equation of the line of regression of x on y is $x = 1.832y - 2.061$. Therefore:

when $y = 2.6$	$x = 2.7022$
when $y = 3.8$	$x = 4.9006$
when $y = 4.9$	$x = 6.9158$
when $y = 8.2$	$x = 12.9614$

    Further:

when $y = 5.5$	$x = 8.015$
when $y = 2.2$	$x = 1.9694$
when $y = 3.4$	$x = 4.1678$
when $y = 4.3$	$x = 5.8166$
when $y = 6.6$	$x = 10.0302$

2. From the data of Example (2) in section 8.5 determine the values of y when x takes the values 2.5, 3.2, 3.6 and 4.5 and determine the values of x when y takes the values 148, 130, 62 and 48.

    The line of regression of y on x is $y = -31.88x + 194.3$. Therefore:

when $x = 2.5$	$y = 114.6$
when $x = 3.2$	$y = 92.284$
when $x = 3.6$	$y = 79.532$
when $x = 4.5$	$y = 50.84$

Further:

$$\begin{aligned} \text{when } x &= 1.5 & y &= 146.48 \\ \text{when } x &= 2.0 & y &= 130.54 \\ \text{when } x &= 2.8 & y &= 105.036 \\ \text{when } x &= 3.7 & y &= 76.344 \\ \text{when } x &= 4.0 & y &= 66.78 \end{aligned}$$

The line of regression of x on y is $x = -0.029y + 5.870$. Therefore:

$$\begin{aligned} \text{when } y &= 148 & x &= 1.578 \\ \text{when } y &= 130 & x &= 2.1 \\ \text{when } y &= 62 & x &= 4.072 \\ \text{when } y &= 48 & x &= 4.478 \end{aligned}$$

Further:

$$\begin{aligned} \text{when } y &= 150 & x &= 1.52 \\ \text{when } y &= 125 & x &= 2.245 \\ \text{when } y &= 91 & x &= 3.231 \\ \text{when } y &= 88 & x &= 3.318 \end{aligned}$$

The interpolated values in the examples above illustrate the points made earlier. For each of the values in the data a prediction has been calculated, using the appropriate regression line. A comparison between the data figures and those of the predictions show how much the error of a prediction can be. The smaller the numerical value of r, the coefficient of correlation, the greater these errors are. Consequently a prediction from a line of regression has to be exercised with tolerance when the degree of correlation is not high.

Exercise 8.2

From the data in each of the three questions below determine the equations of the two lines of regression. From the equations predict values of y and values of x for the related values of x and of y respectively. In each case make an estimate of the degree to which reliance can be placed on the interpolated values.

1.

x:	1	3	5	7	9	11
y:	2.0	4.1	5.9	8.0	10.2	11.9

Determine y when $x = 2, 4, 8, 12, 0.5, 5$ and x when $y = 3.1, 4.7, 6.8, 13.7, 4.1.$

2.

x:	10	15	20	25	30
y:	30.6	27.4	24.5	21.4	18.6

Determine y when $x = 12, 19, 23.2, 26.5, 8, 33.6$ and x when $y = 28.5, 25.2, 20.6, 35.7, 14.8.$

3.

x:	2	4	6	8	10
y:	15.9	21.6	28.3	33.6	40.5

Determine y when $x = 3, 5, 7, 6, 4, 10, 12, 1$ and x when $y = 20.1, 24.7, 30.5, 33.6, 21.6, 10.2, 48.4.$

4. Ten students each sit the same two examinations, A and B. Their marks are tabulated below.

Student:	1	2	3	4	5	6	7	8	9	10
A mark (x):	73	66	48	46	65	60	49	71	74	54
B mark (y):	58	57	21	44	41	61	45	37	66	42

Determine lines of regression of y on x and x on y. Estimate the A marks of students who obtain B marks of 28, 46, 53, 71, 15 and estimate B marks of students who obtain A marks of 63, 52, 83, 29.

ANSWERS TO EXERCISES

To assist students in the solution of certain problems the calculator display value is given as the answer. These values may be rounded off to any degree desired.

Exercise 1.1
1. 1/2   2. 1/2   3. (a) 1/4; (b) 1/4; (c) 1/4
4. (a) 1/2; (b) 1/2; (c) 3/4   5. 3/13   6. 4/13
7. 1/3   8. 1/3   9. 2/3   10. (a) 1/2; (b) 5/6
11. (a) 8/23; (b) 6/23; (c) 9/23   12. (a) n/N; (b) $(n+p)/N$; (c) $[N-(n+p+q)]/N$

Exercise 1.2
1. (a) 4/7; (b) 3/7   2. 1/5; 1/20; 1/12; 1/60; 0; 13/20
3. (a) 5/19; (b) 4/19; (c) 10/19
4. (a) 2/5; (b) 11/23; (c) 5/11; (d) 3/7; (e) 0; (f) 1

Exercise 1.3
1. 1/6   2. 5/6   3. 1/2   4. 1/2   5. 1/3
6. 1/36   7. 1/18   8. 1/12   9. 5/36   10. 1/52
11. 1/2   12. 1/4   13. 1/2   14. 1/8   15. 16/169
16. 1/64   17. $1/52^3 = 1/140608$ 18. (a) 1/24; (b) 3/8; (c) 1/4
19. 4/9   20. (a) 1/16; (b) 1/4; (c) 11/16
21. (a) 16/45; (b) 4/9; (c) 4/45; (d) 7/15
22. (a) 2/55; (b) 49/165; (c) 43/55; (d) 12/55
23. (a) 1/6; (b) 1/3; (c) 2/3   24. (a) 3/32; (b) 119/160
25. (a) 112 749/120 000; (b) 37/30 000
26. (a) 5/72; (b) 1/18; (c) 25/72
27. (a) 0.2925; (b) 0.8075
28. (a) $0.86^4 = 0.5470082$; (b) $0.86^4 + 4(0.86)^3 (0.14) + 6(0.86)^2 (0.14)^2 = 0.9901765$;
    (c) $1 - 0.86^4 = 0.4529918$; (d) $0.14^4 = 0.0003842$
29. (a) 17/35; (b) 11/35; (c) 17/35
30. (a) 33/100; (b) 3/100

Exercise 2.1
1. (a) 1/16; (b) 1/2704; (c) 16/169; (d) 9/169; (e) 9/169; (f) 6/169
2. 41/81; 80/243; 665/729
3. (a) 0.9556716; (b) 0.9999966; (c) 0.0443284

Exercise 2.2
1. $\left(\dfrac{2}{3} + \dfrac{1}{3}t\right)^4$ 2. $\left(\dfrac{5}{6} + \dfrac{1}{6}t\right)^5$ 3. $\left(\dfrac{12}{13} + \dfrac{1}{13}t\right)^4$
4. (a) 1/4; (b) 3/8; (c) 15/16
5. (a) 625/11 664; (b) 31 031/46 656; (c) 64/729; (d) 64/729; (e) 125/15 552

6. (a) $70\left(\dfrac{12}{13}\right)^4\left(\dfrac{1}{13}\right)^4$; (b) $56\left(\dfrac{12}{13}\right)^5\left(\dfrac{1}{13}\right)^3$; (c) $\left(\dfrac{3}{13}\right)^8$; (d) $\left(\dfrac{6}{13}\right)^8$

7. (a) $8\left(\dfrac{1}{4}\right)\left(\dfrac{3}{4}\right)^7$; (b) 58 077/65 536; (c) 649 539/4 194 304; (d) 262 035/262 144

8. (a) 0.0349908; (b) 0.5955988; (c) 0.0565262
9. (a) 0.0595821; (b) 0.3584859; (c) 0.997426

Exercise 2.3
1. 1; 0.980 **2.** 4; 1.549 **3.** 4.8; 1.697
4. 3; 1.449 **5.** 8; 1.265 **6.** 5; 1.581 **7.** 1.5; 1.061
8. 9; 1.225 **9.** 1; 0.949 **10.** 0.1; 0.315

Exercise 2.4
1. (a) 0.147807; (b) 0.9599246; (c) 0.9941987 **2.** 2.160964, i.e. 3
3. (a) 0.0000067; (b) 0.0001535; (c) 0.0114144
4. (a) 0.0000026; (b) 0.2936013; (c) 0.0104064

Exercise 3.1
1. 0.6065307; 0.3032653; 0.0758163; 0.0126361
2. 0.7788008; 0.1947002; 0.0243375; 0.0020281; 0.0001268
3. 0.9048374; 0.0904837; 0.0045242; 0.0000002
4. 0.9900498; 0.0099005; 0.0000495; 0.0000002; 1
5. (a) 0.0000454; (b) 0.000454; (c) 0.00227; (d) 0.12511; (e) 0.6967761
6. (a) 0.9048374; (b) 0.0045242; (c) 0.0001547
7. (a) 0.3678794; (b) 0.7357589; (c) 0.2642411
8. (a) 0.7788008; (b) 0.0021615; (c) 0.9978385

Exercise 3.3
1. (a) 0.1108032; (b) 0.2437669; (c) 0.2681436; (d) 0.6454299; (e) 0.6227137; (f) 0.0249098
2. (a) 0.0001234; (b) 0.8030084; (c) 0.0004394; (d) 0.072765
3. (a) 0.449329; (b) 0.3594632; (c) 0.1437853; (d) 0.0474226
4. (a) 3.08×10^{-29}; (b) 7.0965×10^{-27}; (c) 1
5. 0.6065307; 0.3032653; 0.0758163; 0.0126361; 0.0015795; 0.000158;
(a) 0.0000142; (b) 0.909796
6. At least 195
7. (a) 0.0291134; (b) 0.0004411; (c) 0.9704455

Exercise 4.1
1. 0.5239; 0.6985; 0.9192; 0.9686; 0.9817; 0.99598; 0.99924; 0.99983
2. 0.8180; 0.4010; 0.2846; 0.0818; 0.0188; 0.00328; 0.00804; 0.00152; 0.00034
3. 0.0778; 0.0091; 0.2843; 0.00013 **4.** 0.3859; 0.0516; 0.00307; 0.00052
5. 0.06; 0.24; 0.46; 0.73; 1.83; 2.60; -0.17; -0.33; -0.57; -0.87; -2.91
6. 0.0516 **7.** 0.0084 **8.** 0.3023 **9.** 0.1311
10. 0.14402 **11.** 0.0287 **12.** 0.0102 **13.** 0.4772
14. 0.7745 **15.** 0.14864 **16.** 0.97012 **17.** 0.57
18. 0.11 **19.** 0.32 **20.** -0.05

Exercise 4.2
1. (a) 5; (b) 79; (c) 96; (d) 496; (e) 374
2. (a) 7; (b) 6; (c) 36; (d) 47
3. (a) 0 per cent; (b) 5 per cent; (c) 99 per cent; (d) 100 per cent
4. (a) 0.0951; (b) 0.0055; (c) 0.8994; (d) 0.1006; (e) 0.9243
5. (a) 58; (b) 288; (c) 58; (d) 155; (e) 442
6. (a) 18.75 g $\approx$ 19 g; (b) 0.379 per cent $\approx$ 0.38 per cent; (c) 0.38 per cent; (d) 99.24 per cent
7. (a) 10.56 per cent; (b) 89.44 per cent; (c) 77.34 per cent;
   (d) 4.01 per cent; 3.03 days $\approx$ 3 days

Exercise 4.3
1. Not close to normal, especially at lower and upper limits
2. Fairly close to normal except at lower limit
3. Moderately close to normal
4. Moderately close to normal except at upper limit
5. Extremely close to normal

Exercise 5.1
1. 1.902; 1.648; 1.474; 1.042; 0.851
2. 2.797; 1.978; 1.399; 0.885
3. 3.353; 3.061; 2.834; 2.651; 2.499; 2.371
4. $(\bar{x} - 98.7)/0.0767$; $(\bar{x} - 98.7)/0.0664$; $(\bar{x} - 98.7)/0.0594$; $(\bar{x} - 98.7)/0.042$; $(\bar{x} - 98.7)/0.0343$
5. $(\bar{x} - 202.6)/2.797$; $(\bar{x} - 202.6)/1.978$; $(\bar{x} - 202.6)/1.399$; $(\bar{x} - 202.6)/0.885$
6. $(\bar{x} - 206.5)/3.353$; $(\bar{x} - 206.5)/3.061$; $(\bar{x} - 206.5)/2.834$; $(\bar{x} - 206.5)/2.651$;
   $(\bar{x} - 206.5)/2.499$; $(\bar{x} - 206.5)/2.371$
7. (a) 0.130 per cent or 0.0310 kg; (b) 0.7176; (c) 0.0104
8. (a) 1.699; (b) 0; (c) 1
9. (a) 0.071 mm; (b) 0.1292; (c) 0.3954

Exercise 5.2
1. (a) 34.94 and 32.26; (b) 35.36 and 31.84; (c) 35.85 and 31.35
2. (a) 128.51 and 136.89; (b) 127.19 and 138.21
3. (a) 308.24; (b) 313.03
4. (a) 84.37; (b) 82.89
5. (a) 40.35 and 38.85; (b) 40.74 and 38.46
6. (a) 174.75 cm and 173.25 cm; (b) 173.11 cm; (c) 175.17 cm
7. (a) between 29.2 cm and 29.8 cm; (b) 3078 $\approx$ 3080

Exercise 5.3
1. 375; 49.495   2. 582; 64.322   3. 26.9; 4.4188
4. 106.8; 12.5   5. 1325; 828.68   6. 1325; 847.37
7. 106.8; 13.125   8. 26.9; 4.9
9. 582; 73.143   10. 375; 53.455

Exercise 5.4
1. (a) 40.683 and 34.317; (b) 42.073 and 32.927
2. (a) 112.07 and 101.53; (b) 114.11 and 99.488
3. (a) 4227.5 and 4208.5; (b) 4231.0 and 4205.0
4. (a) 15.134 and 13.146; (b) 15.789 and 12.491
5. (a) 3.787 and 3.249; (b) 3.940 and 3.096
6. (a) 32.980 and 30.995; (b) 33.456 and 30.519

7. (a) 4.083 and 3.659; (b) 4.176 and 3.566
8. (a) 4.042 m/s and 3.980 m/s; (b) 4.056 m/s and 3.966 m/s; (c) 4.076 m/s and 3.946 m/s
9. (a) 44.880 min and 35.453 min; (b) 47.559 min and 32.775 min; (c) 52.760 min and 27.574 min
10. (a) 67.6 and 49.4; (b) 71.6 and 45.4
11. (a) 67.92° F and 62.88° F; (b) 68.82° F and 61.98° F

Exercise 6.1

1. $t = 1.216$; do not accept H_1
2. $t = 1.048$; do not accept H_1
3. $t = 1.824$; at 5 per cent significance level be on the alert for possible increase in the mean
4. $\bar{x} = 5.3695$; $t = 1.686$; do not accept the alternative hypothesis that there has been an increase in the population mean; 99 per cent confidence limits are 5.827 and 4.912
5. $\bar{x} = 27.938$; $s = 3.356$; $t = 1.738$; do not accept the hypothesis that the mean is different from 30. Indeed, do not accept the hypothesis that the mean is less than 30.
6. $t = 10.206$; at the 1 per cent significance level this indicates strong evidence of an increase in the mean breaking strength.
7. Strong evidence at the 1 per cent significance level that the mean is greater than 3.000 cm.

Exercise 6.2

1. $\bar{x}_1 = 404.28$; $\bar{x}_2 = 405.72$
2. $\bar{x}_1 = 53.957$; $\bar{x}_2 = 55.605$
3. $\bar{x}_1 = 139.61$; $\bar{x}_2 = 145.94$
4. $\bar{x}_1 = 471.38$ and 442.22; $\bar{x}_2 = 476.83$ and 436.77
5. $\bar{x}_1 = 77.724$; $\bar{x}_2 = 77.009$
6. $\bar{x}_1 = 2945.7$; $\bar{x}_2 = 2908.5$
7. $\bar{x}_1 = 500.38$; $\bar{x}_2 = 500.55$; when $500.38 < \bar{x} < 500.55$ possible increase in μ; when $500.55 < \bar{x}$ almost certain increase in μ
8. when $500.46 < \bar{x} < 500.62$ or $499.54 > \bar{x} > 499.38$ possible change in μ; when $500.62 < \bar{x}$ or $499.38 > \bar{x}$ almost certain change in μ
9. (a) when $18\,335 < \bar{x} < 18\,488$ possible increase in μ; when $18\,488 < \bar{x}$ almost certain increase in μ
   (b) when $17\,665 > \bar{x} > 17\,512$ possible decrease in μ; when $17\,512 > \bar{x}$ almost certain decrease in μ
   (c) when $18\,404 < \bar{x} < 18\,548$ or $17\,595 > \bar{x} > 17\,452$ possible change in μ; when $18\,548 < \bar{x}$ or $\bar{x} < 17\,452$ almost certain change in μ
10. (a) when $3.161 < \bar{x} < 3.243$ possible increase in μ; when $3.243 < \bar{x}$ almost certain change in μ
    (b) When $2.839 > \bar{x} > 2.758$ possible decrease in μ; when $2.758 > \bar{x}$ almost certain decrease in μ
    (c) when $3.199 < \bar{x} < 3.281$ or $2.801 > \bar{x} > 2.719$ possible change in μ; when $3.281 < \bar{x}$ or $2.719 > \bar{x}$ almost certain change in μ

Exercise 6.3

1. $\bar{x} = 1.3333$; $s = 3.32666$. No evidence even at 5 per cent significance level to suppose that the mean of item A is greater than the mean of item B.
2. $\bar{x} = -13.857$; $s = 9.0079$. Strong evidence at 1 per cent significance level to suppose that the mean of item P is less than the mean of item Q.
3. $\bar{x} = 36.78$; $s = 27.939$. Strong evidence at 1 per cent significance level to suppose that the mean of item X is greater than the mean of item Y.
4. $\bar{x} = 7.6$; $s = 7.9750$. Strong evidence at 1 per cent significance level to suppose that coaching has had an appreciable effect.

Exercise 7.1
1. $\bar{x} = 5.5$; $\bar{y} = 20.1$; $r = 0.9306315$
2. $\bar{p} = 15.1$; $\bar{q} = 19$; $r = 0.9164035$
3. $\bar{U} = 62$; $\bar{V} = 54.9$; $r = 0.4921985$
4. $\bar{x} = 72.5$; $\bar{y} = 10.35$; $r = -0.7363935$
5. $\bar{x} = 35.685714$; $\bar{y} = 1.4757143$; $r = 0.8214387$
6. $\bar{E} = 43.416667$; $\bar{S} = 376.16667$; $r = 0.0775469$
7. $\bar{I} = 106.44444$; $\bar{s} = 2.7555556$; $r = -0.9328364$

Exercise 7.2
1.

n	r_1	r_2	t_1	t_2
5	0.8783113	0.9587362	3.182	5.841
6	0.8113569	0.9171967	2.776	4.604
7	0.7545451	0.8745191	2.571	4.032
8	0.7067471	0.8343124	2.447	3.707
9	0.6664425	0.7976411	2.365	3.499

2. Strong evidence of correlation
3. No evidence of correlation
4. No evidence of correlation
5. Strong evidence of correlation
6. No correlation
7. Strong evidence of correlation
8. $\bar{S} = 1795.6364$ h; $\bar{R} = 391.54545$ cm; $r = -0.7243804$; evidence of possible correlation.

Exercise 8.1
1. $y = 3.3151515\,x + 1.8666667$; $x = 0.2612475\,y + 0.2489254$
2. $p = 1.4143921\,q - 2.3573201$; $q = 0.59375\,p + 3.81875$
3. $V = 0.4744898\,U + 25.481633$; $U = 0.5105682\,V + 33.969805$
4. $y = -0.0139442\,x + 11.360956$; $x = -38.888889\,y + 475$
5. $y = 0.0444917\,x - 0.1120026$; $x = 15.166019\,y + 13.305003$
6. $E = -0.0398193\,S + 63.395367$; $S = -1.9180714\,E + 469.03329$

Exercise 8.2
1. $\bar{y} = 7.0166667$; $\bar{x} = 6$; $r = 0.9995119$;
   $y = 0.9985714\,x + 1.0252381$; $x = 1.0004532\,y - 1.0198469$;
   $y = 3.022381, 5.0195238, 9.0138095, 13.008095, 1.5245238, 6.0180952$;
   $x = 2.0777301, 3.6800191, 5.7830234, 12.692895, 3.0791607$
2. $\bar{y} = 24.5$; $\bar{x} = 20$; $r = -0.9997779$;
   $y = -0.6\,x + 36.5$; $x = -1.6659263\,y + 60.815193$;
   $y = 29.3, 25.1, 22.58, 20.6, 31.7, 16.34$;
   $x = 13.333333, 18.833333, 26.5, 1.3333333, 36.166667$
3. $\bar{y} = 27.98$; $\bar{x} = 6$; $r = 0.9993012$;
   $y = 3.06\,x + 9.62$; $x = 0.3263408\,y - 3.1310162$;
   $y = 18.8, 24.92, 31.04, 27.98, 21.86, 40.22, 46.34, 12.68$;
   $x = 3.4248366, 4.9281046, 6.8235294, 7.8366013, 3.9150327, 0.1895425, 12.673203$
4. $\bar{y} = 47.2$; $\bar{x} = 60.6$; $r = 0.5639021$;
   $y = 0.7053057\,x + 4.4584775$; $x = 0.4508479\,y + 39.31998$;
   $x = 33.37776, 58.89861, 68.823385, 94.344235, 14.946034$;
   $y = 48.892734, 41.134371, 62.998847, 24.912341$

The Normal Distribution Function, (Z)

Z	0.00	0.01	0.02	0.03	0.04	0.05	0.06	0.07	0.08	0.09
0.00	0.5000	0.5040	0.5080	0.5120	0.5160	0.5199	0.5239	0.5279	0.5319	0.5359
0.10	0.5398	0.5438	0.5478	0.5517	0.5557	0.5596	0.5636	0.5675	0.5714	0.5753
0.20	0.5793	0.5832	0.5871	0.5910	0.5948	0.5987	0.6026	0.6064	0.6103	0.6141
0.30	0.6179	0.6217	0.6255	0.6293	0.6331	0.6368	0.6406	0.6443	0.6480	0.6517
0.40	0.6554	0.6591	0.6628	0.6664	0.6700	0.6736	0.6772	0.6808	0.6844	0.6879
0.50	0.6915	0.6950	0.6985	0.7019	0.7054	0.7088	0.7123	0.7157	0.7190	0.7224
0.60	0.7257	0.7291	0.7324	0.7357	0.7389	0.7422	0.7454	0.7486	0.7517	0.7549
0.70	0.7580	0.7611	0.7642	0.7673	0.7704	0.7734	0.7764	0.7794	0.7823	0.7852
0.80	0.7881	0.7910	0.7939	0.7967	0.7995	0.8023	0.8051	0.8078	0.8106	0.8133
0.90	0.8159	0.8186	0.8212	0.8238	0.8264	0.8289	0.8315	0.8340	0.8365	0.8389
1.00	0.8413	0.8438	0.8461	0.8485	0.8508	0.8531	0.8554	0.8577	0.8599	0.8621
1.10	0.8643	0.8665	0.8686	0.8708	0.8729	0.8749	0.8770	0.8790	0.8810	0.8830
1.20	0.8849	0.8869	0.8888	0.8907	0.8925	0.8944	0.8962	0.8980	0.8997	0.9015
1.30	0.9032	0.9049	0.9066	0.9082	0.9099	0.9115	0.9131	0.9147	0.9162	0.9177
1.40	0.9192	0.9207	0.9222	0.9236	0.9251	0.9265	0.9279	0.9292	0.9306	0.9319
1.50	0.9332	0.9345	0.9357	0.9370	0.9382	0.9394	0.9406	0.9418	0.9429	0.9441
1.60	0.9452	0.9463	0.9474	0.9484	0.9495	0.9505	0.9515	0.9525	0.9535	0.9545
1.70	0.9554	0.9564	0.9573	0.9582	0.9591	0.9599	0.9608	0.9616	0.9625	0.9633
1.80	0.9641	0.9649	0.9656	0.9664	0.9671	0.9678	0.9686	0.9693	0.9699	0.9706
1.90	0.9713	0.9719	0.9726	0.9732	0.9738	0.9744	0.9750	0.9756	0.9761	0.9767
2.00	0.97725	0.97778	0.97831	0.97882	0.97932	0.97982	0.98030	0.98077	0.98124	0.98169
2.10	0.98214	0.98257	0.98300	0.98341	0.98382	0.98422	0.98461	0.98500	0.98537	0.98574
2.20	0.98610	0.98645	0.98679	0.98713	0.98745	0.98778	0.98809	0.98840	0.98870	0.98899
2.30	0.98928	0.98956	0.98983	0.99010	0.99036	0.99061	0.99086	0.99111	0.99134	0.99158
2.40	0.99180	0.99202	0.99224	0.99245	0.99266	0.99286	0.99305	0.99324	0.99343	0.99361
2.50	0.99379	0.99396	0.99413	0.99430	0.99446	0.99461	0.99477	0.99492	0.99506	0.99520
2.60	0.99534	0.99547	0.99560	0.99573	0.99585	0.99598	0.99609	0.99621	0.99632	0.99643
2.70	0.99653	0.99664	0.99674	0.99683	0.99693	0.99702	0.99711	0.99720	0.99728	0.99736
2.80	0.99744	0.99752	0.99760	0.99767	0.99774	0.99781	0.99788	0.99795	0.99801	0.99807
2.90	0.99813	0.99819	0.99825	0.99831	0.99836	0.99841	0.99846	0.99851	0.99856	0.99861
3.00	0.99865									
3.1	0.99903									
3.2	0.99931									
3.3	0.99952									
3.4	0.99966									
3.5	0.99977									
3.6	0.99984									
3.7	0.99989									
3.8	0.99993									
3.9	0.99995									
4.0	0.99997									

t Scores, one-tailed test

d.f. \ Pr	.10	.05	.025	.01	.005	.001	.0005
1	3.078	6.314	12.706	31.821	63.657	318.31	636.619
2	1.886	2.920	4.303	6.965	9.925	22.326	31.598
3	1.638	2.353	3.182	4.541	5.841	10.213	12.941
4	1.533	2.132	2.776	3.747	4.604	7.173	8.610
5	1.476	2.015	2.571	3.365	4.032	5.893	6.859
6	1.440	1.943	2.447	3.143	3.707	5.208	5.959
7	1.415	1.895	2.365	2.998	3.499	4.785	5.405
8	1.397	1.860	2.306	2.896	3.355	4.501	5.041
9	1.383	1.833	2.262	2.821	3.250	4.297	4.781
10	1.372	1.812	2.228	2.764	3.169	4.144	4.587
11	1.363	1.796	2.201	2.718	3.106	4.025	4.437
12	1.356	1.782	2.179	2.681	3.055	3.930	4.318
13	1.350	1.771	2.160	2.650	3.012	3.852	4.221
14	1.345	1.761	2.145	2.624	2.977	3.787	4.140
15	1.341	1.753	2.131	2.602	2.947	3.733	4.073
16	1.337	1.746	2.120	2.583	2.921	3.686	4.015
17	1.333	1.740	2.110	2.567	2.898	3.646	3.965
18	1.330	1.734	2.101	2.552	2.878	3.610	3.922
19	1.328	1.729	2.093	2.539	2.861	3.579	3.883
20	1.325	1.725	2.086	2.528	2.845	3.552	3.850
21	1.323	1.721	2.080	2.518	2.831	3.527	3.819
22	1.321	1.717	2.074	2.508	2.819	3.505	3.792
23	1.319	1.714	2.069	2.500	2.807	3.485	3.767
24	1.318	1.711	2.064	2.492	2.797	3.467	3.745
25	1.316	1.708	2.060	2.485	2.787	3.450	3.725
26	1.315	1.706	2.056	2.479	2.779	3.435	3.707
27	1.314	1.703	2.052	2.473	2.771	3.421	3.690
28	1.313	1.701	2.048	2.467	2.763	3.408	3.674
29	1.311	1.699	2.045	2.462	2.756	3.396	3.659
30	1.310	1.697	2.042	2.457	2.750	3.385	3.646

Index